# Drought Management and Desertification Control
## — *Still Miles to Go* —

*About the Centre*

The Centre for Science and Technology of the Non-Aligned and Other Developing Countries (NAM S&T Centre) is an inter-governmental organisation with a membership of 47 countries spread over Asia, Africa, Middle East and Latin America. Besides this, 11 S and T agencies and academic/research institutions of Bolivia, Brazil, India, Nigeria and Turkey are the members of the S and T-Industry Network of the Centre. The Centre was set up in 1989 to promote South-South cooperation through mutually beneficial partnerships among scientists and technologists and scientific organisations in developing countries. It implements a variety of programmes including international workshops, meetings, roundtables, training courses and collaborative projects and brings out scientific publications, including a quarterly Newsletter. It is also implementing 7 Fellowship schemes, namely, Research Training Fellowship for Developing Country Scientists (RTF-DCS), Joint NAM S&T Centre – ZMT Bremen Fellowship, Joint NAM S&T Centre – ICCBS Fellowship, Joint CSIR/CFTRI (Diamond Jubilee) – NAM S&T Centre Fellowship, Joint NAM S&T Centre – DST (South Africa) Training Fellowship on Minerals Processing and Beneficiation, NAM S&T Centre Research Fellowship, NAM S&T Centre – U2ACN2 Research Associateship in Nanosciences and Nanotechnology in Indian institutions. These activities provide, among others, the opportunity for scientist-to-scientist contact and interaction, training and expert assistance, familiarising the scientific community on the latest developments and techniques in the subject areas, and identification of technologies for transfer between member countries. The Centre has so far brought out 79 publications and has organised 108 international workshops and training programmes.

For further details, please visit www.namstct.org or write to the Director General, NAM S&T Centre, Core 6A, 2nd Floor, India Habitat Centre, Lodhi Road, New Delhi-110003, India (Phone: +91-11-24645134/24644974; Fax: +91-11-24644973; E-mail: namstcentre@gmail.com; namstct@bol.net.in).

# Drought Management and Desertification Control
## — *Still Miles to Go* —

*— Editors —*

**Dr R P Dhir**

**Dr Kisamba Mugerwa**

CENTRE FOR SCIENCE & TECHNOLOGY OF THE
NON-ALIGNED AND OTHER DEVELOPING COUNTRIES
(NAM S&T CENTRE)

2018

# Daya Publishing House®

*A Division of*

# Astral International Pvt. Ltd.
New Delhi – 110 002

*Published by*          :  **Daya Publishing House®**
                            *A Division of*
                            **Astral International Pvt. Ltd.**
                            – ISO 9001:2015 Certified Company –
                            4736/23, Ansari Road, Darya Ganj
                            New Delhi-110 002
                            Ph. 011-43549197, 23278134
                            E-mail: info@astralint.com
                            Website: www.astralint.com

*Digitally Printed at*   :  **Replika Press Pvt. Ltd.**

# Preface

It is estimated that land degradation in dry lands, also called desertification, affects ~2 billion hectares area or 52 per cent of all agricultural land globally and that annually some 12 million hectare are lost with a potential loss of 20 million tons of grains. This situation is a threat to livelihood of some 1.5 billion people, most of whom are already poor. Since these are the regions that experience the greatest onslaught also of drought and also that major advances of desertification happen during such periods of climatic vicissitudes, the drought and desertification are inexorably linked. Starvation, loss of livestock, distress migration, accentuation of local conflicts and strife are the common consequences. The people at lower rung of socio-economic strata are the worse affected. Detailed analysis of causes, severity and extent of the problem and of the application of various measures to reverse, or at least prevent further aggravation, have been a focus of concern and action for the past several decades. The experiences in ecological rehabilitation of degraded ecosystems and at minimization of the impact on the affected populations have been quite rewarding. However, it has also thrown up several operational problems: unexpected period of severe drought, inappropriateness of technology from dependent population view point, poor follow up and maintenance, lack of consideration of the nature and magnitude of dependency of the local populations and political instability are the commonly reported impediments in satisfactory functioning. Empowerment and inclusion of weaker sections of the population, peoples' involvement and capacity building are the other key areas that need attention also.

The most durable developments to control desertification and to fight il-effects of droughts have been in situations where natural resource regeneration is accompanied by a reduction in biotic pressure through diversification of rural livelihood enterprises, growth of non-farm employment, expansion of irrigation from endo- or exogenous sources. Small scale industry, cultural tourism, solar power generation, mining, remunerative terms of trade for farmers and livestock breeders. Thus both on-site and off-site measures are essential.

The book comprises papers presented at the "International Workshop on Drought Management and Desertification Control" under auspices NAM S&T and Ferdowsi University of Mashhad. The fourteen papers cover a huge diversity of ecological, land use and management spectrum across the continents. Likewise, the topics cover from just assessment of the land resource degradation problem to a range of experiences at amelioration effort under a common backdrop of a developing economy and complex social structure.

*Dr. R.P. Dhir*

*Dr. Kisamba Mugerwa*

# Introduction

Desertification is land degradation in arid, semi-arid and dry sub-humid areas resulting from various factors, including climatic variations and human activities. This process has already affected one sixth of the world's population, 70 per cent of all dry lands of ~3.6 billion hectares, and one quarter of the total land area of the world. The most obvious impact of desertification, in addition to widespread poverty, is the degradation of 3.3 billion hectares of the total area of rangeland, constituting 73 per cent of the rangeland with a low potential for human and animal carrying capacity; decline in soil fertility and soil structure on about 47 per cent of the dryland areas constituting marginal rain fed cropland; and the degradation of irrigated cropland, amounting to 30 per cent of the dry land areas with a high population density and agricultural potential. The priority in combating desertification involves implementing the preventive measures for lands that have still not degraded, or are only slightly degraded. However, the severely degraded areas also need to be kept in focus. In combating desertification and drought, the participation of local communities, rural agencies, national governments, non-governmental organisations (NGOs) and international and regional organisations is essential.

Keeping the above in view, the NAM S&T Centre in partnership with the Ferdowsi University of Mashhad organised an International Workshop on 'Drought Management and Desertification Control' during 22-24 May 2017 at Ferdowsi University, Mashhad, Iran which brought various stake holders, *viz.* scientists, experts and professionals engaged in R&D, policy making and implementation, to a common platform for up gradation of their skills and sharing views and experiences in the drought management and desertification control.

The Mashhad International Workshop was attended by 72 scientists, experts and professionals from 15 developing countries, including Bangladesh, Bhutan, Cuba,

India, Indonesia, Iraq, Malaysia, Mauritius, Myanmar, Nepal, Nigeria, Palestine, Sri Lanka, Uganda and the host country Iran.

As a follow up of the above workshop, the present book – 'Drought Management and Desertification Control- Still Miles to Go' - has been edited by Dr. R.P. Dhir and Dr. Kisamba Mugerwa . There are 14 scientific technical papers contributed by the experts from 11 countries.

The publication is an outcome of the highly laudable involvement and valuable efforts of the entire team of the NAM S&T Centre, especially of Dr. Kavita Mehra, Mr. M. Bandyopadhyay, Ms. Rashmi Srivastava and Ms. Geeta at all the stages of the publication process. The contribution of Mr. Pankaj Buttan in designing of the cover page, formatting and liaising with the printers is worthy of mention.

I am sure that this book will be useful to all those associated in combating desertification and drought for sustainable food production including the researchers, local communities, rural agencies, national governments, non-governmental organisations (NGOs) and regional and international organisations.

*Prof. Dr. Arun P. Kulshreshtha*
*Director General,*
*NAM S&T Centre*

# Contents

*Chapter 1*

# Is Groundwater Table Depleting in Barind Tracts Over Time?

*Md. Shafiqul Islam[1], Mahbuba Nasreen[2] and Hamidul Huq[3]*

[1]*Doctoral Researcher and Assistant Professor,*
*Center for Sustainable Development, University of Liberal Arts, Bangladesh*
*E-mail: shafiqul.islam@ulab.edu.bd*
[2]*Professor and Director,*
*Institute of Disaster Management and Vulnerability Studies, University of Dhaka*
[3]*Professor, Department of Economics,*
*United International University, Bangladesh*

## Abstract

The study has been conducted to know the groundwater status in the Barind Tracts in Bangladesh. Groundwater has been depleted over times in Barind in Bangladesh. Ten years data for six wells from six locations have been studied and analyzed for the study. Rainfed agriculture has been replaced by the irrigated agriculture. Around eighty per cent agricultural land is now cultivated through groundwater irrigation. Barind Multipurpose Development Authority (BMDA) has the active contribution in irrigated agriculture, but groundwater table is falling down consecutively all over the Barind Tracts due to over-withdrawal. The groundwater depletion is a great shock in northwestern Barind due to its geographical settings. The groundwater has been recharged by rainfalls only. The rainfall pattern in the study area is erratic and lowest within the country. After analyzing data, the hydrographs represent the status of groundwater in the study areas. These show that groundwater table is depleting over time parallel with increase in irrigated agriculture. Information from interviews with focus group supports this finding also. Agricultural production has been increased through irrigation and cropping intensity in the study area, but groundwater table is declining at the rate of 2.028 ft/y in dry season and 1.892 ft/y in wet season. Rainfed agriculture, agroforestry, organic agriculture, crop diversification, use of surface water, creation of surface water reservoirs, increasing irrigation efficiency and rainwater harvesting are the possible options in improving the situation.

*Keywords: Drought, Groundwater depletion, Irrigation, Organic agriculture, Rain fed agriculture.*

## Introduction

Northwest region of Bangladesh covers Rajshahi, Chapai Nawabganj and Naogaon districts. The area is dominated by a tropical moist downpour climate. Bangladesh is the land of Agriculture and rice is the staple food. Rice alone contributes almost eighty per cent to the total food supply. Rice farming in Bangladesh differs with the change of season and water supply. There are distinct three rice growing seasons in Bangladesh, likely January to June (Boro), April to August (Aus), and August to December (Aman) wrapping roughly 11.0 million hectares of land in Bangladesh (Awal and Siddique, 2011). The contribution of Boro rice was 56.47 per cent and 55.77 per cent in total rice production in Bangladesh during the period of 2009-10 and 2010-11 respectively (BBS, 2011). Eventually dry season Boro rice production mostly relies on additional irrigation and rain fed agriculture has been displaced by irrigated agriculture. Drought is risky to crops as they grow depending on rainfall during the monsoon. Due to lack of surface water in dry season, groundwater substitutes the irrigation water causing relentless groundwater reduction and may be an immense warning for future groundwater availability for Boro rice production especially in Barind Tracts of Bangladesh.

The Northern Barind of Bangladesh is currently experiencing water paucity troubles in agriculture and safe livelihood (Alice, 2010). Monsoon rainfalls and flooding contributes to groundwater recharging in Bangladesh. Barind is situated in the highly elevated and flood free zone. Rainfall is recharging groundwater in this Barind, erratic and low rainfall happens in northwestern part of Bangladesh and the area has turned into extremely drought prone area. Furthermore, the Barind Tract soil types which are thick, sticky clay obstruct groundwater recharging and enhance surface runoff. Drought may affect groundwater in both dry and wet seasons. The groundwater recharge is also affected if there are droughts in the monsoon season. This affects both the surface and groundwater situation negatively and the irrigation in the following dry seasons (Choudhury *et al.*, 2005). Water shortfalls vary from 400–500 mm in the dry periods. The river flow of the Mohananda and the Punarbhava tends to decline during the dry season. Groundwater declination rate has been increased after crop intensification through tube well irrigation which began in the early 1980s. The Groundwater table in High Barind Tracts is depleting continuously over the time especially in dry season due to over exploitation of groundwater by tube well (Islam *et al.*, 2010). The groundwater level in this area is consecutively declining with escalating extraction of water for irrigation (Rahman and Mahbub, 2012). Careful monitoring of groundwater level for agriculture is essential.There is an immediate need of study on irrigated crops and groundwater table suction limit.

## Literature Review

This study examines One fourth of world's irrigated lands is provided by groundwater, of which 75 per cent of lands are to be found in Asia (Shamsudduha *et al.*, 2011). Bangladesh Agriculture was completely reliant on nature including surface water and monsoon rainfall inearly 1970s (UNDP, 1982). In Bangladesh, 79.1 per cent of lands are irrigated using undergroundwater during Boro season (BADC, 2010). Bangladesh Agriculture was led and irrigated by traditional means up to 1950s. The Barind Tract is situated in the north- western part of Bangladesh with total area of 7727 square kilometers (Rasheed, 2008). Physically, this Barind Tract stands between 24°20′ N and 25°35′N latitudes and 88° 20′E and 89° 30′E longitudes. Barind Tract was built up by Pleistocene Alluvium also known as Mature Alluvium and grounded by reddish brown, muggy Pleistocene deposit; Madhupur mud (Ahmed, 2006). Barind Tract was debarred all through 3000 Deep Tube Well (DTW) installation programme of BADC in North-west Irrigation Project, allowing for low impending area for groundwater development (BMDA,2006). Groundwater development takes place in Barind by creation of Barind Integrated Area Development Project (BIADP) in 1985 under BADC and afterward through creation of Barind Multi- purpose Development Authority (BMDA) in 1992 (BMDA, 2011). Groundwater reduction rate in dry and wet seasons is diverse. Groundwater reduction rate in wet season is higher than the rate in dry season. Study findings reveal that groundwater level is depleting in an accelerating rate over the time (Rahman and Mahbub, 2012). It also needed to find out suitable and vulnerable area of irrigation. The geological condition of an area governs the occurrence and distribution of groundwater. The groundwater recharge is also affected if there are droughts in the monsoon season. This affects both the surface and groundwater situation negatively the irrigation in the following dry seasons (Choudhury *et al.*, 2005). Water shortfalls vary from 400–500 mm in the dry periods. The river flow of the Mohananda and the Punarbhava leans to decline during dry season. Groundwater declination rate has been increased after crop intensification through tube well irrigation began in the early 1980s. The Groundwater table in High Barind Tracts is depleting continuously over the time especially in dry season due to over exploitation of groundwater by tube well (Islam *et al.*, 2010). The groundwater level in this area is consecutively declining over times with escalating extraction of water for irrigation (Rahman and Mahbub, 2012).

The study was conducted with the view to know whether groundwater table is depleting or not in the Northwest Barind Tracts. The other associate objectives of this study are to:

⭑ Understand the groundwater situation in the study location.

⭑ Determine the seasonal fluctuation of groundwater in different locations over time

⭑ Know the comparative trend of groundwater depletion in the study location

⭑ Predict future trend of groundwater table using regression equation for next 10 years

✯ The future recommendation of rice production with Groundwater fluctuation and variation.

The study also tested the following hypothesis: "There is a significance declining trend of groundwater table among the locations over time"

## Methodology

The study was pedestal on groundwater monitoring wells data. The data was obtained from Barind Multipurpose Development Authority (BMDA) for the period of 2005-2014. Secondary data of six wells from six mouza (each from Upazila) were collected from Barind Multipurpose Development Authority (BMDA). Paramanadapurmouza from Godagari, Haripurmouza from Tanore, Chandpurmouza from Shibganj, Somaspurmouza from Nachole, Darajpurmouza from Niamatpur and Balashohidmouza from Porsha were selected for data. Six locations were selected from six Upazilas from three districts. These selections were done based on drought severity ranking (source CEIGS, 2013). Depth was calculated in feet. Automatic avometer was used in recording SWL data from wells. Total 343 respondents were interviewed using random sampling and semi structured questionnaire. Additionally, individual interviews with experts and Focus Group Discussions (FGDs) were carried out for this study. Hydrographs were prepared for this study to make the status visualize. Actually time sequence data are not bivariate data; simple regression analysis technique was used to draw linear trend line (Udofia, 2004, Okoko, 21001). In this study, the independent variable is time in years (x) and Static groundwater table for 10 years (2005-2014) is dependent variable (y). The least square model is presented as;

$$Y = a + bx + e$$

*where,*

Y = Dependent variable (Static groundwater level in feet)

X = Independent variable (time in years).

a = A constant and y – intercept

b = Regression coefficient

c = Error random term

Time in years was the only considered cause of the fluctuations as the meteorological controls of the annual rainfall vary in their positions and intensities periodically.

The chi-square tests of relationship wherever there are sets of variables for comparison are in use for it determines homogeneity of baseline indices. The test of contingency, k–sample chi-square test of homogeneity is used to relate the rainfall, SWL descriptive indices with the study locations. The study was to elicit that there is significant variation in the indices having similar milieu. The problem can be solved as a 5 x 4 contingency problem utilizing the rather normal chi-square test formula. The use of the conventional $x^2$ formula involves the calculation of the expected frequencies using the formula.

$$Fe = FtrFtc/N$$

*where,*

Fe = Expected frequency

Ftr = Total row frequency

Ftc = Total column frequency

N = Total frequency

The complete elements of the usual chi-square test are given below as;

$$X^2 = (O-E)^2/E$$

*where,*

O = Observed frequency

E = Expected frequency

n = Number of categories

## Results and Discussion

### Irrigated Agriculture and Drought

The analysis of static groundwater level indicates the present status of groundwater with the future predicted trend. Rice farming in Bangladesh differs with the change of season and water supply. There are distinct three rice growing seasons in Bangladesh, likely Boro (January to June), Aus (April to August), and Aman (August to December) wrapping roughly 11.0 million hectares land in Bangladesh (Awal and Siddique, 2011). The contribution of Boro rice was 56.47 per cent and 55.77 per cent in total rice production in Bangladesh during the period of 2009-10 and 2010-11 respectively (BBS, 2011). Eventually in a dry season Boro rice production mostly relies on additional irrigation. Drought is risky to crops as they grow dependent on rainfall during the monsoon. Due to lack of surface water in a dry season, groundwater substitutes the irrigation water causing relentless groundwater reduction and may be an immense warning for future groundwater availability for Boro rice production especially in Barind Tracts of Bangladesh. The Northern Barind of Bangladesh is currently experiencing water paucity troubles in agriculture and safe livelihood (Alice, 2010)]. Monsoon rainfalls and flooding contributes in groundwater recharging in Bangladesh. Barind is situated in the highly elevated and flood free zone. Rainfall is recharging groundwater in the Barind area, as erratic and low rainfall happens in northwestern part of Bangladesh and the area has turned into extremely drought prone area. Furthermore, the soil types (thick sticky clay) of Barind Tract obstruct groundwater recharging and enhance surface runoff.

### How Groundwater Decline by Drought and Irrigated Agriculture

The groundwater recharge is also affected if there is drought in the monsoon season. This affects both the surface and groundwater situation vis-à-vis the irrigation in the following dry seasons (Choudhury *et al.*, 2005). Water shortfalls vary from 400–500 mm in the dry periods. The river flow of the Mohananda and the Punarbhava tends to decline during dry season. Groundwater declination rate has been increased after crop intensification since tube well irrigation began in the early 1980s. The Groundwater table in High Barind Tracts is depleting continuously over the time especially in dry season due to over exploitation of groundwater by tube well (Islam *et al.*, 2011). The groundwater level in this area is consecutively declining over time with escalating extraction of water for irrigation (Rahman and Mahbub, 2012). Careful monitoring of groundwater level for agriculture is essential, as there is an immediate need of study on better crop yield and groundwater table suction limit. Currently, one fourth of world's irrigated lands is provided by groundwater, of which 75 per cent of lands are to be found in Asia (Shamsudduha *et al.*, 2011). Bangladesh agriculture was completely reliant on nature including surface water and monsoon rainfall in the early 1970s (UNDP, 1982). In Bangladesh, 79.1 per cent of the lands are irrigated using undergroundwater during Boro season (BADC, 2010). Bangladesh agriculture was irrigated by traditional means up to 1950s.

## Barind Multipurpose Development Authority and Irrigated Agriculture

The Barind Tract is situated in the north- western part of Bangladesh with total area of 7727 square kilometer (Rasheed, 2008). Physically, this Barind Tract stands between 24/20'N and 25/35'N latitudes and 88/20'E and 89/30'E longitudes. Barind Tract built up by Pleistocene Alluvium also known as Mature Alluvium and grounded by reddish brown, muggy Pleistocene deposit; Madhupur mud (Ahmed, 2006). Barind Tract was debarred all through 3000 Deep Tube Well (DTW) installation programme of BADC in North-west Irrigation Project allowing for as low impending area for groundwater development (BMDA,2006). Groundwater development takes place in Barind by creation of Barind Integrated Area Development Project (BIADP) in 1985 under BADC and afterwards through the creation of Barind Multi-purpose Development Authority (BMDA) in 1992 (BMDA, 2011). Groundwater reduction rate in dry and wet seasons is diverse. Groundwater reduction rate in wet season is higher than the rate in dry season. Study findings reveals that groundwater level in the study area is depleting in an accelerating rate over the time (Rahman and Mahbub, 2012). Corresponding Table 1.1 shows the maximum (100.78 per cent) irrigated lands were increased at NacholeUpazila over times with maximum depletion of groundwater (3.39 feet/yr) and increased numbers (180) of Deep tube well. The maximum numbers of deep tube well was increased in Nachole followed by Godagari, Niamatpur, Porsha and Tanore respectively. It was observed that the numbers of deep tube well was decreased at Shibganj Upazila.

## The State of Groundwater as Perceived by the Local People

During the discussion with the Focus Group participants, most of them mentioned that groundwater table was going down over time. The pump and shallow tube well does not work due to lack of groundwater table. Lot of tube well and pumps are now out of work. Even they reported that huge groundwater was used after intervention of BMDA irrigation project.

**Table 1.1: Irrigated Area, Number of Deep Tube Well and Fluctuation Rate of Static Groundwater Level**

| Description | Locations | | | | | |
|---|---|---|---|---|---|---|
| | *Niamat-pur* | *Nachole* | *Goda-gari* | *Tanore* | *Shibganj* | *Porsha* |
| Irrigated land in 2014 (Hectares) | 36915 | 42213 | 64949 | 38717 | 8655 | 14010 |
| Irrigated land in 2005 (Hectares) | 26415 | 21025 | 37176 | 33299 | 7280 | 7390 |
| Difference in Hectares | 10500 | 21188 | 27773 | 5418 | 1375 | 6620 |
| Percentage increase in Irrigated land | 39.75 | 100.78 | 74.71 | 16.27 | 18.89 | 89.58 |
| Groundwater depletion in feet/Yr | **2.75** | **3.39** | **2.06** | **1.76** | 0.2 | 1.69 |
| No. of deep tube well in 2014 | 660 | 544 | 714 | 552 | 238 | 283 |
| No. of deep tube well in 2005 | 522 | 364 | 558 | 489 | 245 | 192 |
| Increased/decreased number of deep tube well | **138** | **180** | **156** | **63** | **-7** | **91** |

*Source*: BMDA data (2015).

## Sign of Groundwater Table Depletion

Several signs were picked up during the interview and focus group discussions in relation to groundwater depletion. The signs included lack of water availability in the tube well, water crisis, high temperature, less or no rainfall, drying of pond, canal and rivers. Most of the respondents (n=285) marked that tube well did not

**Table 1.2: Indication of Groundwater Table Depletion**

| Causes | Study Location with Respondents (Number and percentage) | | | | | | All N=343 |
|---|---|---|---|---|---|---|---|
| | *Rajshahi* | | *Chapainawabganj* | | *Naogaon* | | |
| | *Parisho n=52* | *Aye-Hi n=62* | *Nizampur n=75* | *Chokghor-pakhia n=51* | *Bhabicha n=63* | *Mollapara n=40* | |
| No water from tube well | 52 (100) | 62 (100) | 74 (98.66) | 44 (86.24) | 33 (52.38) | 20 (50) | 285 (83.09) |
| Lack of water | | | | | 9 (14.29) | 2 (5) | 11 (3.21) |
| High temperature | | 1 (1.61) | | | 9 (14.29) | 2 (5) | 12 (3.50) |
| Less rainfall | | 2 (3.23) | | | | 3(7.5) | 5 (1.46) |
| Drying of pond | 1 (1.92) | 11 (17.7) | | | 13 (20.63) | | 25 (7.29) |
| Drying of river | | | | | 1(1.59) | | 1 (0.29) |

*Source*: Field data (2015) and multiple responses.

work as the sign of groundwater depletion (Table 1.2). Actually the main sign of groundwater depletion is all types of tube well and pumps go out of work or perform partially.

## Analysis of Water Level Hydrographs

By using ten years data, groundwater level hydrographs were prepared based on fortnightly data collected from monitoring wells of BMDA from 2005 to 2014. Only one monitoring well was considered for each Upazila. The fluctuation of groundwater depth was shown by the hydrographs. It was observed that the mean water table in Godagari Upazila was 63.9 feet with the maximum 85.8 feet (December 2014) and minimum 30.2 feet (July 2006). Corresponding figure for Tanore Upazila the maximum groundwater table was 53.7 feet (April 2009) and minimum level was 21.8 feet (July 2005) with the average 38.3 feet (Figure 1.1). Shibganj Upazila showed less fluctuation comparing other locations. The hydrographs in Shibganj is more or less very gentle in nature. The average water table was 16.2 feet with the minimum groundwater table 6.2 feet in the month of September, 2008 and maximum 27.7 feet in the month of March, 2011 (Figure 1.2). In Nachole, the lowest groundwater table was 49.7 feet in the month of January, 2006 and maximum groundwater table was in the month of April, 2014 (109.4 feet) with average 92.2 feet (Figure 1.2). Niamatpur

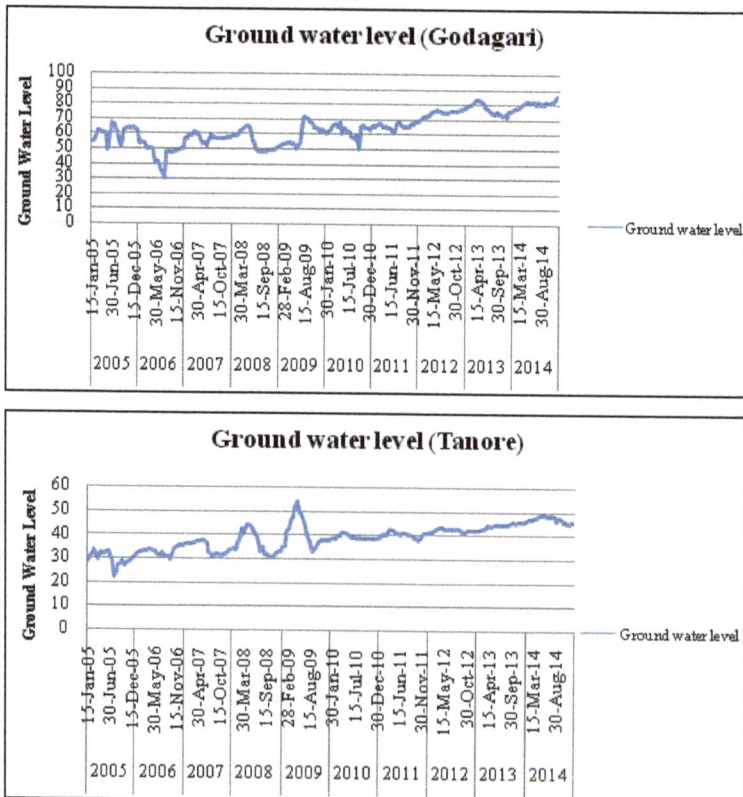

**Figure 1.1: Hydrographs for Godagari and Tanore.**

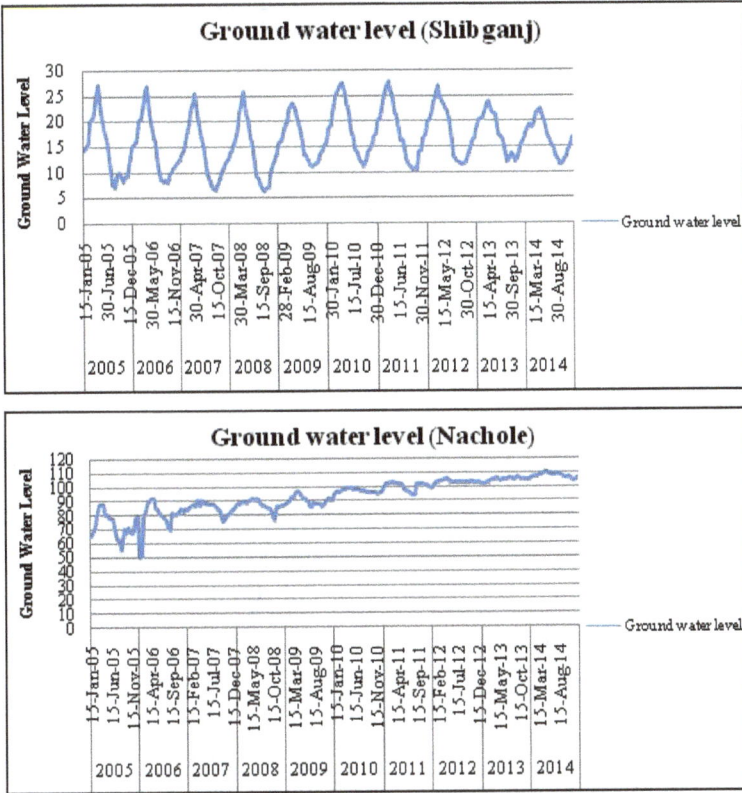

Figure 1.2: Hydrographs for Shibganj and Nachole.

represented the lowest groundwater table in the month of December 2005 (14 feet) with the highest groundwater table in the month of April,2014 (58.3 feet) with the average level of 39.3 feet. On the other hand, the average groundwater table in Porsha was 17.9 feet. The maximum groundwater table was 58.3 feet (March, 2014) and minimum was 1.8 feet in the month of January, 2006 (Figure 1.3). The hydrographs of all Upazilas except Shibganj, show that groundwater depth is declining day by day but the significant levels started from 2010 in most of the Upazilas.

## Groundwater Restraints

Ten years data on static water table (SWL) from six different locations (six wells) were collected for the period of 2005 to 2014. All six locations were considered for the calculation of average SWL for 2014 and 2005 and eventually dry and wet seasons. Analyzing six wells data, it was observed that groundwater is depleting over the time. The maximum annual depletion rate was found in Nachole followed by Niamatpur, Godagari, Tanore, Porsha and Shibganj respectively (Table 1.3). The maximum depletion rate was 3.39 feet in a year with the minimum depletion rate was 0.2 feet (Shibganj). The rate of recharging is too much lower considering withdrawal

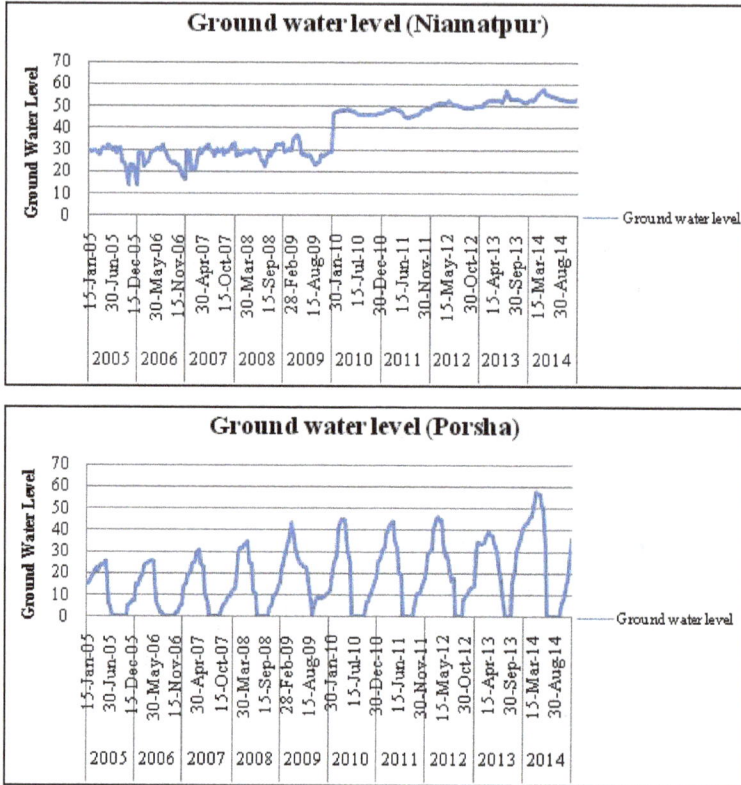

**Figure 1.3: Hydrographs for Niamatpur and Porsha.**

of groundwater in the study area. The water table continued at the maximum depth during dry season and minimum during wet season. The groundwater table is diminishing in shocking rate in the study area of Barind Tract with the expansion of irrigation scheme.

**Table 1.3: Average and Fluctuation Rate of Static Groundwater Level in Feet**

| Description | Locations | | | | | |
|---|---|---|---|---|---|---|
| | Niamatpur | Nachole | Godagari | Tanore | Shibganj | Porsha |
| Static Groundwater level in 2014 (Feet) | 54.4 | 106.8 | 80.85 | 46.7 | 16.7 | 27.2 |
| Static Groundwater level in 2005 (Feet) | 27 | 72.8 | 60.23 | 29.1 | 14.7 | 10.3 |
| Difference in Feet | 27.5 | 33.9 | 20.62 | 17.6 | 2 | 16.9 |
| Time period (2014-2005) 10yr | 10 | 10 | 10 | 10 | 10 | 10 |
| Fluctuation in Feet/Yr | 2.75 | 3.39 | 2.06 | 1.76 | 0.2 | 1.69 |

*Source*: BMDA data (2015).

## Groundwater Restraint in Dry Season

January to May was considered as the dry season due to no or less rainfall pattern.The groundwater depletion rate was calculated based on average value of the data for five months from January to May and it was then put in the table. Corresponding Table 1.3, only shibganj showed no depletion of groundwater during dry season. The maximum groundwater depletion was found in Porsha (2.95 fee per year) followed by Nachole (2.93 feet/yr), Niamatpur (2.53 feet/yr), Godagari (2.16 feet/yr) and Tanore (1.63 feet/yr) respectively (Table 1.4). Water table drops and attains at the maximum depth from surface and the height continued minimum from the sea level during dry season. The minimum depletion rate of groundwater in dry season was found in the Shibganj (Chandpur) may perhaps be due to closeness of location from the river and soil types which give maximum recharge and higher water retention capacity.

**Table 1.4: Average and Fluctuation Rate of Static Groundwater Level in Feet during Dry Season (January to May)**

| Description | Locations | | | | | |
|---|---|---|---|---|---|---|
| | Niamatpur | Nachole | Godagari | Tanore | Shibganj | Porsha |
| Static Groundwater level in 2014 (Feet) | 55.3 | 107.4 | 79.44 | 47.3 | 19.96 | 50.6 |
| Static Groundwater level in 2005 (Feet) | 30 | 78.1 | 57.89 | 30.9 | 20.23 | 21.1 |
| Difference in Feet | 25.3 | 29.3 | 21.55 | 16.3 | -0.27 | 29.5 |
| Time period (2014-2005) 10yr | 10 | 10 | 10 | 10 | 10 | 10 |
| Fluctuation in Feet/Yr | 2.53 | 2.93 | 2.16 | 1.63 | -0.027 | 2.95 |

*Source*: BMDA data (2015).

## Groundwater Restraint in Wet Season

The wet season was considered from June to September. Normally rain occurs during wet season in Bangladesh. The maximum 3.96 feet/yr groundwater depletion was recorded at Nachole during wet season followed by Niamatpur (2.75feet/yr), Godagari (1.973feet/yr), Tanore (1.97feet/yr), Porsha (0.36feet/yr) and Shibganj (0.342 feet/yr) respectively (Table 1.5). The depletion rate of groundwater between wet and dry season differs from locations to locations. The lowest depletion rate of groundwater was found in wet season due to availability of precipitation. Groundwater recharging occurs in rainy season in connecting the month of June to October. Almost eighty per cent rainfall occurs in monsoon period in Bangladesh (Rahman and Mahbub, 2012). The deficit was created by excess extraction of groundwater over recharge. According to Rahman and Mahbub (2012), annual maximum rate of depletion and minimum rate of depletion in TanoreUpazila from five monitoring wells (two different organizations BWDB and BMDA) extend over diverse parts of the study area is 1.04feet/year.

**Table 1.5: Average and Fluctuation Rate of Static Groundwater Level in Feet during Wet Season (June to September)**

| Description | Locations | | | | | |
|---|---|---|---|---|---|---|
| | *Niamatpur* | *Nachole* | *Godagari* | *Tanore* | *Shibganj* | *Porsha* |
| Static Groundwater level in 2014 (Feet) | 54.1 | 107 | 81.11 | 46.9 | 14.12 | 4.7 |
| Static Groundwater level in 2005 (Feet) | 26.6 | 67.4 | 61.38 | 27.3 | 10.70 | 1.1 |
| Difference in Feet | 27.5 | 39.6 | 19.73 | 19.7 | 3.42 | 3.6 |
| Time period (2014-2005) 10yr | 10 | 10 | 10 | 10 | 10 | 10 |
| Fluctuation in Feet/Yr | 2.75 | 3.96 | 1.973 | 1.97 | 0.342 | 0.36 |

*Source*: BMDA data (2015).

## Analysis of Groundwater Trend

The trend of groundwater level was observed using regression equation for different locations from the study area. Six wells data from six different locations were considered for this study. Ten years long term trends of groundwater level of Godagari, Tanore, Shibganj, Nachole,Niamatpur and Porsha were considered. It was observed that in all locations groundwater table is going down (Figures 1.4–1.9).

## Predicted Static Groundwater Level using Regression Equation (2016-2025)

Using regression equation for six study locations, the predicted groundwater table was calculated and it was found that the maximum groundwater depletion was

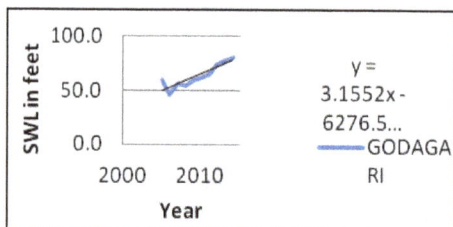

Figure 1.4: Groundwater Trends in Godagari.

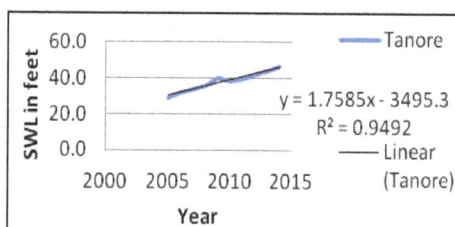

Figure 1.5: Groundwater Trends in Tanore.

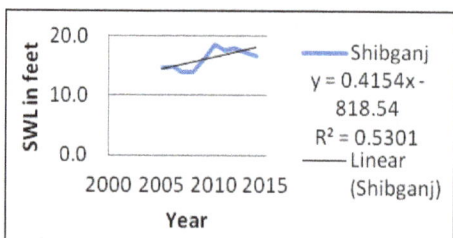

Figure 1.6: Groundwater Trends in Shibganj.

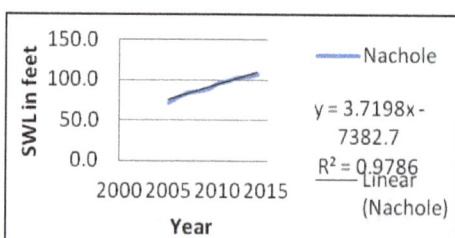

Figure 1.7: Groundwater Trends in Nachole.

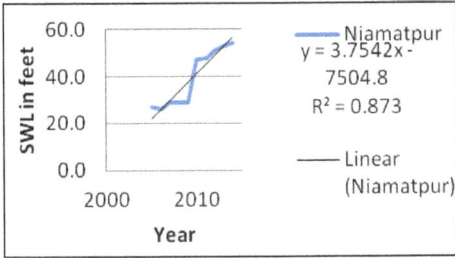

**Figure 1.8: Groundwater Trends in Niamatpur.**

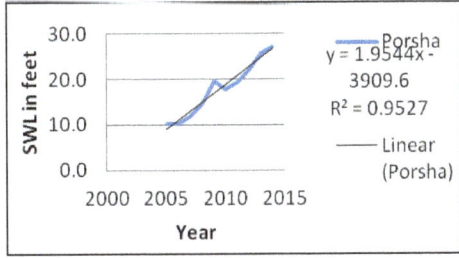

**Figure 1.9: Groundwater Trends in Porsha.**

found in Niamatpur followed by Nachole, Godagari, Porsha, Tanore and Shibganj respectively over following next 10 years (Table 1.6).

**Table 1.6: Predicted Static Groundwater Table for the Period of 2016-2025**

| Year | Location | | | | | |
|------|----------|---------|----------|--------|----------|--------|
|      | Niamatpur | Nachole | Godagari | Tanore | Shibganj | Porsha |
| 2016 | 64.1 | 115.5 | 84.5 | 49.1 | 18.1 | 30.3 |
| 2017 | 67.8 | 119.2 | 87.6 | 50.9 | 18.6 | 32.2 |
| 2018 | 71.6 | 122.9 | 90.8 | 52.6 | 19.0 | 34.2 |
| 2019 | 75.3 | 126.7 | 93.9 | 54.4 | 19.4 | 36.1 |
| 2020 | 79.1 | 130.4 | 97.1 | 56.2 | 19.8 | 38.1 |
| 2021 | 82.8 | 134.1 | 100.3 | 57.9 | 20.2 | 40.0 |
| 2022 | 86.6 | 137.8 | 103.4 | 59.7 | 20.6 | 42.0 |
| 2023 | 90.3 | 141.5 | 106.6 | 61.4 | 21.0 | 43.9 |
| 2024 | 94.1 | 145.3 | 109.7 | 63.2 | 21.5 | 45.9 |
| 2025 | 97.9 | 149.0 | 112.9 | 64.9 | 21.9 | 47.8 |

## Testing of Hypothesis

The static groundwater level data was used to derive descriptive statistics of mean, standard deviation, coefficient of variation with maximum and minimum value for the periods of 2005 to 2014 (Table 1.7).

**Table 1.7: Static Groundwater Level Descriptive Indices for Six Study Locations in Barind Tracts (2005-2014)**

| Determinants | Godagari | Tanore | Shibganj | Nachole | Niamatpur | Porsha |
|--------------|----------|--------|----------|---------|-----------|--------|
| Mean | 63.9 | 38.3 | 16.2 | 92.2 | 39.3 | 17.9 |
| Maximum | 80.9 | 46.7 | 18.6 | 106.8 | 54.4 | 27.2 |
| Minimum | 46.8 | 29.1 | 14.7 | 72.8 | 26.2 | 12.0 |
| SD | 10.7 | 5.5 | 1.7 | 11.4 | 12.2 | 6.1 |
| CV | 107.2 | 54.6 | 17.3 | 113.8 | 121.6 | 60.6 |

The expected frequencies (Fe) were calculated using the determination for each entry in Table 2 and 3, the test statistic is appended below:

$X^{22} = (O-E)^2/E$ is utilized for static groundwater level

$$\frac{(63.93-72.5)^2}{72.5} + \cdots\cdots + \frac{(17.9-62.2)^2}{62.2} + \frac{(80.9-77.2)^2}{77.2}$$

$$+ \cdots\cdots + \frac{(27.2-66.23)^2}{66.23} + \frac{(46.8-25.1)^2}{25.1} + \cdots\cdots + \frac{(12-21.51)^2}{21.51}$$

$$+ \frac{(10.7-12.4)^2}{12.4} + \cdots\cdots + \frac{(6.1-10.68)^2}{10.68} + \frac{(107.2-39.4)^2}{39.4}$$

$$+ \cdots\cdots + \frac{(60.6-33.82)^2}{33.82} = 190.7$$

The level of confidence (99 per cent) and (number of rows-1) (number of columns-1) = (5-1) (6-1) = 20 degrees of freedom, the calculated chi-square value *i.e.* calculated value is greater than critical value (190.7 › 37.5662). The null hypothesis was denied and hence static groundwater level decreased significantly among the locations over times. It was found that groundwater is depleting due to drought and huge withdrawal of groundwater for irrigation in the study area. All of the locations except Shibganj are very much critical. The areas are usually recharging by rainfalls which is also declining over the periods.

## Conclusions

Groundwater table is depleting in the study area due to huge withdrawal of groundwater and surface water over time. River flows and rainfalls are the main sources of recharging groundwater. Another hindrance of groundwater recharge is poor infiltration capacity of the soil in the Barind Tracts. As a result, quick run off occurs and discharges huge rain water during monsoon. Average depletion rates in the dry season is 2.028 feet/year and depletion rates in wet season is 1.892 feet/year respectively. Rate of declination of groundwater table in dry season is higher than that of wet season which means groundwater recharge coming down due to withdrawal of excessive groundwater for irrigated crops. It can be said that Upazilas of Nachole, Niamatpur, Porsha, Tanore and Godagari are very much vulnerable for irrigation. Filling of ponds, siltation of water body and rivers are leading low water reserve for surface irrigation with the ultimate result of groundwater depletion. There is direct relation between groundwater and irrigated crops especially Boro *i.e.* irrigated rice production and groundwater depletion. So we need to change the cropping pattern instead of irrigated crops. Hence, few environmental negative effects may arise like excessive temperature, reduced surface water supplies, and erratic or less rainfalls. Diversification of crops should be done from more water consuming crop (paddy) to less water consuming crops (vegetables, fruits *etc.*), excavation and re-excavation of ponds, canal, *khal, beel,* cultivation of drought resistant crops (Seasam, mung), agroforestry, organic agriculture, increasing dependency on surface water, irrigation efficiency application of Alternate Wetting and Drying (AWD) method, rainwater harvesting *etc.*, can be options for the study area. It is also important to find out suitable and vulnerable area of irrigation. So, evaluation is needed to measure the geological condition in this region.

# References

1. M. Ali, "Fundamentals of Irrigation and On-farm Water Management," Springer-Verlag New York Inc., New York, Vol. 1, 2011, p. 18.

2. BADC (Bangladesh Agricultural Development Corpora-tion), "Minor Irrigation Survey Report 2009-10," BADC, Dhaka, 2010, p. 3.

3. BBS (Bangladesh Bureau of Statistics), ⁻Yearbook of Agricultural Statistics of Bangladesh ,2011.

4. BMDA (Barind Multipurpose Development Authority), "Borandro Authority Past-Present," BMDA Rajshahi, 2006, p. 35.

5. BMDA (Barind Multipurpose Development Authority), 2011. http: //www. bmda.gov.bd/

6. Choudhury, A. M., Neelormi, S., Quadir, D. A., Mallick, S., and Ahmed, A. U. (2005). Socio-economic and physical perspectives of water related vulnerability to climate change: results of field study in Bangladesh. *Science and Culture*, 71(7/8), 225.

7. Islam, M. B., Ali, M. Y., Amin, M., and Zaman, S. M. (2010).*climatic variations: farming systems and livelihoods in the high barind tract and coastal areas of Bangladesh* (pp. 477-497). Springer Netherlands.

8. K. B. S. Rasheed, "Bangladesh: Resource and Environ-mental Profile," A.H. Development Publishing House, Dhaka, 2008.

9. K. M. Ahmed, "Barind Tract,"2006. http: //www.banglapedia.org/httpdocs/ HT/B_0309.HTM

10. M. Alice,Research Report on Water Scarcity in Northern Bangladesh, 2010

11. M.A. Awal, M.A.B. Siddique, Rice Production in Bangladesh Employing By Arima Model, *Bangladesh J. Agril. Res.* 36(1),pp 51-52, 2011.

12. M.M. Rahman,A.Q.M. Mahbub,⁻Groundwater Depletion with Expansion of Irrigation in Barind Tract: A Case Study of TanoreUpazila,Journal of Water Resource and Protection.pp. 567-575, 2012.

13. M. Shamsudduha, R. G. Taylor, K. M. Ahmed and A. Zahid, "The Impact of Intensive Groundwater Abstraction on Recharge to a Shallow Regional Aquifer System: Evidence from Bangladesh."

14. Hydrogeology Journal, Vol. 19, No. 4, 2011, pp. 901-916. doi: 10.1007/s10040-011-0723-4

15. S.K. Adhikary, A.A. Sharif,S.K. Das, G.C. Saha, ⁻Geostatistical Analysis of Groundwater

16. Level Fluctuations in the Shallow Aquifer of Northwestern Bangladesh, Proceedings of the 2nd International Conference on Civil Engineering for Sustainable Development,14~16 February,2014, KUET, Khulna, Bangladesh, p. 391.

17. Okoko,E. (2000). Quantitative techniques in urban analysis. Ibadan, Krafy Books

18. Udofia, E.P.(2008). Fundamentals of social science statistics, Enugu, Immaculate Books.

19. UNDP (United Nation Development Programme), "Ground- water Survey: The Hydro-Geological Conditions of Bang-ladesh," Technical Report, UNDP, New York, 1982, DP/UN/BGD-74-009/1.

*Chapter 2*

# Desertification Perspectives: A Case Study of Indian Thar Desert

*R.P. Dhir*

*Formerly Principal Scientist and Director,*
*Central Arid Zone Research Institute,*
*ICAR, Jodhpur, India*
*E-mail: dhirrp08@gmail.com*

## Abstract

Desertification has remained on the international agenda for the past more than four decades. As a result immense amount of work has been done across the affected countries to understand the forms of desertification, to map the severity and extent of the problem and to elaborate the ways it affects the integrity of cultural and natural ecosystems or impacts the livelihood of the people depending upon these. Though importance of social dimension of desertification has been realized, application of the same in planning and implementation of control programs has left much to be desired. Since several geographic regions experiencing desertification are beset also with under-development, political issues and unfair social disparities, the application and maintenance of remedial have their own shortcomings. Land degradation has several areas of convergence with conventions like those on climate change and biological diversity but there are also areas of conflict. Over all, the problem of desertification, despite its social, economic and environmental impacts, remains under-addressed.

Experiences over the past have shown the positive role of interventions that are not directly part of desertification control. Measures such as poverty alleviation and improvements in literacy, public health, and infrastructure and service sector growth allow the affected populations a better ability to live with the problem of land degradation but help in several ways in restoration activity. Thar Desert has seen many such transformations amongst which, irrigation development has reduced dependence on rainfed farming and in the process it has also stabilized the landforms that were once threatening the well being of the people. Public food distribution system has practically done away with possible starvation and deprivation. Besides developments in rural infrastructure, growth of non-farm employment, rural housing has helped in minimizing pressure on land. China and few other countries have achieved similar results also.

*Keywords: Desertification and development, Socio-economic dimension, Thar experience.*

## Introduction

Increase in human and livestock pressure has progressively led to increased land degradation, the severity of the problem and its impact being particularly grievous in developing countries. The semi-arid and arid areas, where desertification is most widespread because of the fragility of natural and cultural ecosystems, are also known for frequent occurrence of drought and such periods of high rainfall deficit, besides severely impacting the grain and biomass production of the given year, pave way for an accelerated advancement of land degradation processes. The drift sands originating from anthropogenic activities and a strong wind regime encroach upon fertile lands in vicinity. There has been national and international effort to comprehend the forms and processes of desertification and its causative factors. Over time major emphasis came to be laid on ways and means to reduce, if not totally reverse, the severity of problem. There has been a growing concern on socio-economic dimension of the problem both as an agent of resource degradation and of development.

The present paper traces history of the rather huge amount of thinking and action that have taken place from local to international levels. The situation of desertification and development of northwest arid zone, popularly called Thar, has been discussed in detail to show development, application and maintenance issues of various well-meaning and technically sound technological interventions in situation of high biotic pressure.

## Concept of Desertification and International Framework

The problem of desertification came on international agenda in preparation of and during the UN Conference on Desertification in 1977. Some nations and independent scientific experts from the world over presented their views on the nature of the problem, causative factors and possible technologies and approaches to cope up with the situation. Important follow ups were the definition of the word 'desertification" and the effort for regional and global assessments of the problem. The definition that came to be adopted in course of time extended the geographic coverage from arid zone that lay in the immediate vicinity of the already existing deserts and were most vulnerable and/or were getting desertified rapidly, to semi-arid and dry sub-humid regions of the world. This definition was based on the recognition that the problem of land degradation and its impacts were equally serious or even greater in the other dry lands. Following the Sahelian situation of those times, the droughts were considered to abate or greatly accelerate the pace in several ways and hence climate variability or drought was incorporated in the definition as well. Importantly, preparation of action plans at national or regional levels and an implementation of the same with domestic and/or international support was a major outcome also.

The period that followed saw several reviews, consultations and regional and international experience-sharing workshops or conferences, the high point of which was the Rio Earth Summit in 1992. This activity provided a major platform also for a new dimension that arose out of the realization that various facets of desertification were an outcome of exploitative human activities and that success can be achieved only through awareness and the concerted actions involving the concerns of local people. In fact people were considered major stakeholders, whose interests and concerns ought to prime in all degradation control activities. The other major outcome of the Earth Summit was the resolution by of the United Nations General Assembly to establish an Intergovernmental Negotiating Committee (INCD) for preparation of a Convention to Combat Desertification (UNCCD), which entered into force on 26 December 1996. A permanent secretariat for the Convention for co-ordination and promotion of various international activities in form of UNCCD was established in 1997 in Germany and the Global Environment Facility (GEF) was its financial mechanism. Till to date, several 'Conferences of the Parties' have been organized where national action plans and future programs are presented besides an overview of UNCCD activities, including the 10-year strategic plan and framework to enhance the competency of control effort.

The above-said Rio Earth Summit promoted also two other conventions namely the UN Framework Convention on Climate Change (UNFCCC) and the UN Convention on Biological Diversity. Both of these also got ratified and are in force for more than two decades. During the meetings, several momentous decisions have followed including the Kyoto Protocol, some of which, like the greenhouse gas emission level norms, have become binding under the international law. Desertification Convention, the UNFCCCs and UNBD in several ways have a common goal to rejuvenate or protect environment and in turn ensure livelihoods.

## Brief Review of Approach and Desertification Control Efforts in Developing Countries

Despite decades of interventions, desertification, drought and land degradation (DDLD) have remained in focus at local, national and international levels. Several integrated or activity-focused programs have been undertaken world over. The extent, severity and consequences of land degradation and desertification as well as effectiveness and sustainability of control efforts have been discussed. An attempt is made to briefly describe these.

### Approaches to Desertification Control

A number of processes alone or in combination such as wind and water erosion, vegetation degradation, salinization, loss of water resources and diminution of land productivity cause desertification and scientific research has provided technologies to ameliorate the situation. However, within a few years of application of these activities, it became increasingly clear that desired results were not forthcoming in several cases. It was primarily due to the fact that either the measures adopted did not prove effective enough or that these could not be sustained in the prevailing driving forces that caused the degradation in the first place. Soil and water conservation treatments mostly involved earth work structures that had a limited life span. Further, an uncommon spell of heavy rainfall was enough to undo much of the effort. The households with subsistence level existence had little resources or machinery to repair and re-operationalize the structures. The developed grazing lands needed a regulated grazing for continued productivity but the high livestock load and inadequacies in community organization did not allow a management regime in the prevailing open access situations. The problem of maintenance was further compounded by the socio-economic inequalities (Mortimore, 2003). The resource-poor amongst the dependent populations normally are the worst affected by the consequences of degradation and they are also the ones, who for bare survival, cause overgrazing, cutting of trees for fuel or make exploitative use of the small chunk of agricultural land that they possess (DDC, 2003). The development plans in the prevailing social and economic order did not provide incentives or other preferential capacity-building activities for them. Political instability or social conflicts such as between settled and migratory populations came in the way also of a coherent policy and programme of implementation (Eriksen, 2003; Darkoh, 2003).

Furthermore, the people living in these marginally suited environments have evolved through experience over the generations a variety of indigenous management systems. Though some of these surely had become out of date in new means of production and livelihood systems, several had a merit and relevance. But neglect of these systems and livelihood strategies reduced the acceptance or workability of the practices introduced in the new development paradigm (Eriksen, 2003). Besides, the physical or biological interventions were made to conserve natural resources, which often enough did not take note of the huge dependence that the local people had on these for their very existence. Spooner (1982) had elaborated quite early the social perspectives of desertification that were put forcefully later at the 1992 Rio Earth Summit document also. These latter comprised topics like combating poverty, sustainablity enhancement of local populations and human resource development in general. Since then a lot of advocacy has taken place to make desertification control program people-centric in its approach, planning and implementation. But the latter is easier said than done since an accelerated rehabilitation of degraded resources does call for a restraint on prevailing exploitative use of resources that are being developed. Mortimore (2003) has focused on several development and maintenance ambiguities that prevail in the complex human-environment systems in dry lands and that are poorly understood, such as alignment of development policy with dry land peoples' own strategies, learning from success or failure of development exercises and lack of understanding of linkages between and within diverse groups.

As several case studies have shown the problem of land degradation is most serious in countries or their sub-regions that are saddled also with a poor economy whereas the steps to check desertification and rehabilitating degraded lands are expensive and time-consuming. Further, continuance of natural resource degradation accentuates poverty still more by greater exploitation of whatever of natural resources is left. Thus a vicious cycle is set in motion and earlier this is broken, better it is. But for ill-developed, desertification-prone regions, rejuvenation effort can happen from larger national resources or with external assistance. However, financial allocations have varied across regions and over time but have been generally short of the need. It is also a fact that indigenous enthusiasm, proper use of funds and lack of infrastructure have left much to be desired. The high variability in climate parameters, particularly in rainfall and its distribution, constrain the success of the effort and this comes in the way of regular flow of external assistance, despite a call from several international agencies for liberal support in the cause of environment protection and reducing misery of the people and their societies. Even small incentives such as providing a drinking water storage structure, a feed trough for livestock, thrashing floor or addition of diversity in local enterprise have enhanced motivation and interest in the big task of natural resource regeneration but the role of these remains underappreciated (Mortimore, 2003) as also that of the social-desertification linkages as a whole (Smith and Reynolds, 2003).

As mentioned earlier, individual countries have been working on control of desertification and land degradation for a longtime. Impressive gains have been made in areas of afforestation, shelter-belts, soil and water conservation, sand dune stabilization and control of salinization. However, critical, independent reviews are few regarding the success of these efforts or of the extent to which the advancement of land degradation problem has been minimized. But there is a widespread consensus that the problem has been inadequately addressed (Low, 2013) and the reasons vary from structural, operational, economic to political.

## Assessment of Severity and Extent of Desertification

Estimates on the magnitude of a problem are basic to planning, allocation of resources and implementation of restorative measures. Therefore, rightly an activity for data generation on forms and spatial distribution of desertification was launched right from the beginning. An outcome of this was the generation of statistics and maps in form of an atlas that was first printed in 1992 and second edition appeared in 1997 (Middleton and Thomas, 1997). These and later updated information showed that worldwide the land degradation and desertification were a serious problem on ~2 billion hectares area and besides, this showed also that 52 per cent of all agricultural land was moderately to severely affected. At the same time the problem was worsening and each year ~12 million hectare are lost with a potential loss of 20 million tons of grains (Low, 2013). This situation is a threat to livelihood of some 1.5 billion people, most of whom were already poor. Besides the above, several countries have made detailed or holistic studies also with use of tools like remote sensing. Regarding India for example such studies show that nearly half of the country is affected by one or the other form of degradation, within which water erosion, wind erosion, water logging, soil salinity and soil acidity constitute 63.9, 6.3, 9.7, 4.0, 10.9 per cent respectively (Samra and Sharma, 2005; Anonymous, 2010).

However, reviewers or those who scrutinized it as well as the experts engaged in generation of information were aware of the paucity of hard data base regarding severity and type of degradation. Water erosion is a major process of degradation but it has immense spatial variation even across agricultural fields within a village. The small scale maps or remote sensing data that were generated thus showed only a broad generalization and varied depending upon the nature of experience and scientific background of the survey or the ground knowledge and proficiency of the remote sensing specialists. Darkoh (2003) has shown how assessments have varied amongst different agencies for the same region.

Besides, several observations have been made on the very premises that underlie the concept of desertification. One of these observations is that continuation of land degradation in long run leads to virtual loss of land productivity and that leads ultimately to desert-like situation. But experience over the years has not borne out this assumption. For example, the grazing lands in most dry lands under continuous overuse do lose their productivity but even after long periods of time all it means is that the these lands acquire a low-level equilibrium of production (Dhir, 1995). Another case is of sandy soils that dominate dry lands in several parts of the world, particularly in Asia, Africa and Middle East. Because of their high basic erodibility, these soils invariably experience wind erosion depending upon the wind regime as the associated aridity does not allow adoption of sufficient protective vegetative measures. The problem is common both to degraded grazing lands as well as to cropped lands. But despite a long duration of an exploitative management, a ruinous loss of productivity has not been observed (Dhir, 1995; Warren and Olsson, 2003). There is some recuperation in soil following a return to a period of milder wind erosion or an affordable extra effort of the farmers at land leveling and some protection. Likewise, removal of most fertile components of soil namely fine particles and organic matter is a major cause of loss of productivity potential in wind or water eroded lands. However, the consequences in terms of loss of plant nutrients leading to infertility of soils are really serious in shallow soils only. Furthermore, dry lands farmers make effort to apply silt and other compensatory mechanisms to improve the soil. Several of these common experiences and management practices are often ignored in assessing implications of soil degradation. Going further, the environmental impacts are not the sole outcome of physical changes but are tied to the ability of a society or culture to adapt to those changes as also to the coping ability of individual farmers. Mortimore (2003) and Darkoh (2003) have highlighted several such ambiguities in functioning of complex human-environment relationships. Summing up, though the problem exists, there has been some loss of credibility of the term 'desertification or degradation' because of oversimplifications and several generalizations (Dhir, 1999; Eriksen, 2003).

Human population, through activities related to land use and management, is a major driving force in desertification. An increase in its pressure impacts through increased need for more food and feed and reduction in size of land holdings. Both amount to an intensification of land use as well as expansion of cropping into areas that are marginal in nature. Thus, population growth has been considered all along an accelerator of the desertification problem. However, there have been regions where developments in economy such as urbanization, industrialization, mining of bountiful minerals and fossil fuels, infrastructure have been not only able to absorb the load of surplus population but have been helpful in reducing dependence on traditional livelihoods. Jiang (2002) cites the case of Inner Mongolia, where the human population has tripled in four decades but the income per capita through various developments, including irrigation, has increased fifty fold. Thus, the relationship of population growth to desertification is not simple and depends on collateral and independent economic developments (Darkoh, 2003).

## Drought and Desertification

Droughts are part of normal climate regime of a given region and people living particularly in marginal environments are forced to borrow money, make distress sale of livestock, face malnutrition or even starvation and undergo temporary migration with all the attendant miseries. Over time with increased pressure on land and or due to the effect of ongoing land degradation, the impact of droughts on livelihoods has worsened. Numerous studies have shown a drastic decrease in land productivity. This is also the period when incidence of overgrazing is at its maximum and the surviving cover of vegetation, particularly grasses, suffers the most. Rightly so, diminution of ecosystem resilience and major advancement of desertification happens in years with continuation of drought situation for two or more years in succession. Therefore, justifiably drought or climate variability are strongly linked with land degradation and these impinge also on the success of rehabilitation effort. However, experience has shown that the impact of degradation varies depending upon the national capacity to extend relief and minimize rigors of migration. Therefore, at the same resilience of ecosystem, adaptive capacity of the management systems and economic strength of the affected populations have been considered important (Smith and Renold, 2003; Bradley and Grainger,2004).

## Economics of Desertification

There has been a quest always to know what land degradation means in economic terms and it gained a greater necessity to know the cost of this process and benefits that will flow from it in order to satisfy the needs of policy makers and donors. Thus based on consultations with experts, including Dregne (1983), UNEP came out with an estimate in 1980's that the direct cost of desertification at global level was $26 billion per year. The costs were classified also as direct (decrease in crop area, reduction in crop yields, poor response to input use, and decline in profitability of enterprise), indirect (siltation of reservoirs, effect on public health, costs in infrastructure maintenance and others); and economy-wide (impact on national development goals). At local levels a huge amount of data on the effect of degradation on crop yields exists, though as said above, the yields over time are determined not just by desertification but also by several other factors. Furthermore, efforts continue to be made to assess the costs at various scales and for various regions (Dregne *et al.*, 1991; Bojo, 1996). A large amount of data has been generated for China. Cheng *et al.* (2013) show that the average direct and indirect cost of degradation, though varied amongst the studies made, was ~1 per cent of GDP. In India, Reddy (2003) showed that direct costs of land degradation in various States varied mostly from 0.2 to 1.9 per cent of GDP with a national average of 0.89 per cent. Such studies exist for various parts of the dry lands, though thoroughness and inadequacy of scientific and economic data do leave much to be desired.

Regarding benefits of control effort, the information is even more inadequate for reasons such as that the interventions were more for social or environmental objectives and the income was not easily measurable or that the gestation period was long, as for example for afforestation or grassland development. However, in most cases the benefits outweigh the cost involved, particularly so in areas like control of waterlogging, salt-affected soil or acid soil reclamation and plant nutrient applications. Fleskens *et al.* (2012) developed a model "Desertification Mitigation Cost Effectiveness and showed that cost-benefit analysis varied across situations depending on the responsiveness to technology. In their study they found that use of a technology across cereal growing area was profitable only in one-third of the applicable area (Fleskens *et al.*, 2013). Therefore, in marginally suited areas, attention needs to be given to subsidies since gains can be marginal as compared to the cost of restorative effort. Besides the above, globalization of trade and inability of weak economies to have fair trade terms have affected land degradation. The price differential in agricultural and industrial products constrains the economy of the developing world. Rise in demand and hence the prices lead to overstocking and accelerate desertification in several central African countries (Rocheleu *et al.*, 1995).

## Desertification and Development

As has been said above in several contexts, dry lands across the world are poorly developed in terms of infrastructure, communications, non-farm economy, and industry, partly because of low asset and economic base of the people making a living therein and partly because of natural disadvantages manifest in situations like water scarcity, dusty atmosphere and inadequate skill or competency level of workers. However, in situations where such handicapped regions form a part of much larger geographic area and economy, situation is different. Despite their inability to contribute, such drylands receive benefits of investments in key areas. As described later in this paper, Thar Desert has seen a lot of development in irrigation, part of it from externally sourced water. This activity alone through high value and increased intensity of cropping has been a sizeable factor that has benefitted households and the region as a whole in several ways. Greater employment opportunities and diversification have been the other spinoff's with much needed softening the impact of drought. Jiang (2002) has shown how in Inner Mongolia, even when the human population has multiplied three-times, the cash income of people has jumped fifty folds. The above is only to show that socio-economic effort in terms poverty reduction, growth of infrastructure, industry and urbanization, mining and service sectors can impact through reduced dependence on primary industry and exploitation of natural resources.

## Desertification, Climate Change and Biodiversity Degradation

Besides a convention on desertification, the outcome of the 1992 Earth Summit was launching of two more conventions, namely the UN Framework Convention on Climate Change (UNFCCC) and the UN Convention on Biological Diversity (UNCBD). All the three have functioned all along and efforts have been made to identify areas of convergence between them. For example, global warming is seen to increase climate variability, make rainfall more stormy and increase water need of crops as well as of vegetation, the desertification is seen to further worsen the impact of climate change. On the other hand changes in land use and management practices to store and sequester carbon to slow global warming are integral to control of desertification and to an extent also to bio-diversity build up (Stringer *et al.,* 2012). Whereas conservation of biodiversity and promotion of carbon sequestration require development of untouched natural ecosystems, these are unacceptable to mixed farmers and pastoralists, for whom utilization of the same ecosystem is a mainstay for their livestock, fuel, fencing and timber. This conflict of interest is particularly serious in productivity-wise marginal regions. For example, for climate change animals are methane producing machines whereas for local populations through domestic animals these are the basic means of making a livelihood. The same applies to freshwater and marine fishing activities that are almost the sole means of livelihoods for landless people in coastal regions. Thus a major agenda for the informed people, experts and governments is to find ways to enhance the convergence with minimum damage to main goals.

## Thar Desert: A Case of Desertification and Development

Thar Desert, climatically speaking, is an arid zone with an annual rainfall in range of 100 to 450 mm as against a potential evapo-transpiration value of 1700 to 2400 mm. It distinguishes itself by a monsoonal pattern of rainfall but with a high inter-annual variation in amount and distribution (CV 44-64). It has an area of 0.2 million square kilometer, of which nearly two-thirds are made up of sandy plains and dunes (Singh, 1982). The region is fairly rich in biodiversity and much of it is endemic. There are over seven hundred species of trees, shrubs and herbs, most of these are generally hardy, have an ability to make most of rainfall and are multi-purpose in their utility. Perennial grass species are amongst the renowned species for their outstanding quality. But the natural ecosystem today is highly transformed due to rather intense human activities. Understandably, surface and groundwater resources are scare but the region has some respectable aquifers, the storage of which is mainly from favourable climate periods in the remote past. With a publication record of over 5,000 in number, the arid zone has an impressive record of scientific investigation, technology development and extension effort (Faroda, 1998 and Kar, *et al.,* 2009).

## History of Human Settlement, Population Growth and Land Use

Like several other regions in Asia, Thar has a long history of human settlement, the pre-historic era being dominated by tribes, who were food gatherers and hunters, though they maintained domestic animals. Even though, the Indus/Saraswati Civilization flourished in the westernmost part of Thar, it had not penetrated into the Thar proper. Influx of modern populations into this region started both from the south in Gujarat and the plains in the north in early historic period. But population in the TharDesert all along was thin with animal husbandry as the main pursuit. Crop husbandry then was confined to climatically better half eastern part with considerable irrigated agriculture along drainage lines issuing from the Aravalli mountains in the east. However, human population growth was sluggish and some studies have shown that the numbers increased only three-fold in 230 year period to year 1890 (Dhir, 1982). The situation of sluggish growth continued to census year 1921.This situation was mainly due to high maternal and infant mortality rate, famines and epidemics like cholera, malaria and typhoid. However, from then on the growth has been continuous and big such that the population tripled in fifty years to 1971 and tripled again in forty years to year 2011. With a density of 135 persons for every square kilometer, TharDesert has become the most densely populated arid zones in the world.

Like human population, arable farming has put up a large growth in recent years. Whereas in 1930s the cropped area was 10-30 per cent across much of the desert, it had risen to 36.3 per cent of the entire region in year 1956 and to 45.7 and 57.4 per cent respectively in years 1980 and 2010. Much of this growth was at the expanse of reduction in incidence of fallow farming and by breaking new lands in areas considered marginal climate-wise. Other main categories of land use are culturable wastelands, unculturable waste lands which constitute 17.6 and 4.6 per cent respectively. Designated grazing lands constitute 3.8 per cent Only but the just said waste lands are used also for grazing purposes.

## Agriculture, Animal Husbandry and Traditional Management Practices

With 50-60 per cent of population dependent on it, agriculture is by far the major livelihood source of the people. Because of recent developments, as of today of the total area of ~12 million hectare under cultivation, 74 per cent is rainfed and the rest has the benefit of irrigation. Of the latter little more than half is from groundwater and the rest from canal irrigation, the water source of which is the Himalayan rivers outside the Thar. Under rainfed farming,pearlmillet, pulses and oil seeds are the crops. In a specialized situation with some amount of winter rainfall, gram is grown also as a rainfed crop.

Over time, the cropping pattern has shown some commercial dimension with a large increase in area of cluster bean. Some medicinal plants have come to occupy a significant area also. This commercial dimension is even stronger under irrigation with groundnut, cotton, mustard and a variety of condiments and spices, fodder and fruit trees as the crops. Animals are an important source of economy. Presently these number over 30 million, *i.e.* more than the size of human population, with cattle, buffalo, goat, sheep and camel constituting 21, 13, 23, 42 and 1 per cent respectively. The present composition is an outcome of a significant change, under which the share of goat, sheep and buffalo have increased considerably.

Local people over the generations had developed several management and strategies and most of these remain as relevant today as they were in the past, particularly in area of rainfed farming and in livestock rearing. Agro-forestry and mixed cropping with a combination of short and medium duration crops to cope with uncertainties of rainfall amount and distribution are specifically conspicuous. The crops namely millet, pulses and clusterbean are hardy, generally with short duration and yet they have an ability to take advantage of a good rainfall year. The crops are also multipurpose with food grains and fodder being important outputs. Fallow farming, crop rotations, choice of crops in relation to onset of rain are the other highlights. Farming based on rainwater harvesting has been an old practice in the driest part of Thar.

In area of animal husbandry mixed flock has been a strategy to make best use of multi-storey vegetation. Breeding practices, ethno-veterinary, migration are the other well developed practices. The change in livestock composition in recent years in response to ecological compulsions or market forces shows that livestock rearing is viable. In regard to the total population, nearly 2/3$^{rd}$ is presently with farmers, the rest being with the pastoralists. The mixed farmers find animal component an asset as they are able to make use of the residual biomass and give in turn milk and manure, prolong duration of gainful employment, minimize the adverse effect of drought and lead to women empowerment as most of the animal rearing operations are carried out by women.

## Drivers and Manifestations of Desertification

Past historical evidence suggests that natural resource degradation was not a serious problem as to pose a problem in human settlement and making of a living save for periods of extended drought. Furthermore, few situations that have remained under pristine situation with very restricted human activity till to date show that landscape is well vegetated and landforms quite stable even in the driest part of Thar (so (Figure 2.1). Hence, all the various manifestations so abundant presently are an outcome of human activity, mainly of the past century or so.

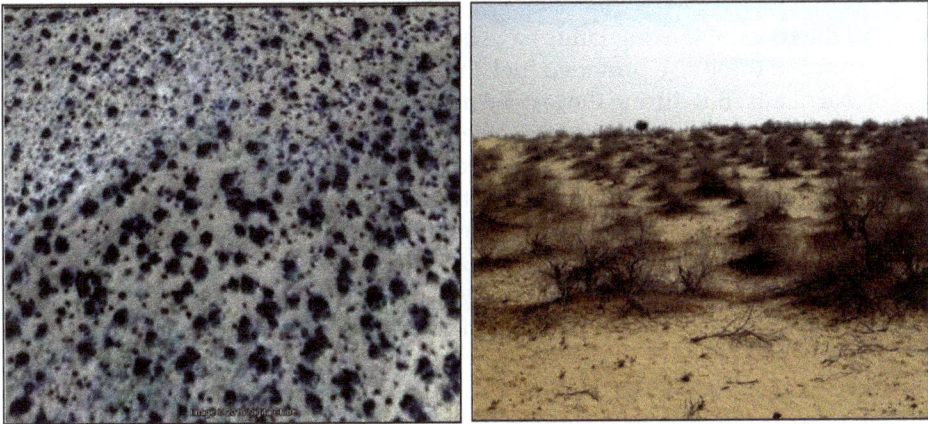

Figure 2.1: (a) Satellite View of Natural Vegetation Cover on Duny Landscape in the Driest Part *i.e.* with ~100 mm Mean Annual Rainfall, and (b) A Field View of the Same.The situation is as much a tribute to adaptability of natural vegetation as it is to moisture conservation property of sand soil.

In comparison, similar landform even in the rainfall-wise better situation, has very sparse cover and possesses abundant manifestation of instability. Thus the degradation is essentially due to exploitative management of natural resource base. Human activities impact in several ways. A major driving force has been an expansion of crop husbandry activity. This has happened in two ways; firstly through its expansion to the drier western part, which was earlier considered marginal for this purpose. The other has been intensification of this activity in the eastern part at the cost of earlier fallow-farming. In the latter area even somewhat saline, rocky, gravelly or shallow soil tracts have been brought under plough. These developments have been an outcome of increased human population pressure and tractorization of land preparation and sowing operations.

Earlier, bullock-drawn systems were in capable of putting an entire holding of a farmer under crop because of the limited period for which surface soil remained moist for seeds to be sown, but tractor offered no limitation in this regard. Even a bigger effect of this was the deep ploughing using tractor-drawn disc. This practice was found helpful in turning over of the soil and in minimizing the problem of weeds, which is otherwise is quite serious. But this greatly enhanced the incidence of wind erosion. An equally adverse effect of this practice has been experienced on the prized agro-forestry system. Shrubs, like zizyphus, which is hardy, produces feed of outstanding quality were numerous in agricultural fields but because of tractor cultivation the shrub stand has suffered hugely and the present productivity is just $1/4^{th}$ to $1/10^{th}$ of what it was in the past. The stand of the multipurpose tree, the "khejri" (*Prosopis cinerarias*), has suffered a great deal also. The accelerated wind erosion causes generation of masses of drift sands that pile up against obstacles and field boundary or creates shrub-coppice dunes or hummocks that disturb the level of the land and necessitate a leveling operation. Furthermore, the loose sand comes in the way of obtaining a uniform and adequate crop plant population.

All the types of grazing lands have been the main plank of animal husbandry in the past and these also provided fuel, thatch and other useful biomass, including medicinal plants. But during the last 5-6 decades the area of these lands has shrunk because of the above-said expansion of cropping. But more grievous has been the depletion of useful vegetation cover because of persistent overgrazing (Suresh Kumar, 1997). This degradation manifests through adverse change in species number and cover of climax vegetation type and progressively in loss of vigor and ultimately in their disappearance (Saxena, 1977; Shankarnarayan, 1988). Since grass species are a major component of arid zone ecosystem and are also the preferred grazable matter, perennial grasses are worst affected and in the process these are replaced by annual grasses and ephemeral herbs with a greatly reduced quantity of grazable biomass (Saxena, 1977, 1993; Shankar and Kumar, 1988). These degraded lands are invaded also by unpalatable shrubs such as *Calotropis* sp., *Aerva* sp. and *Haloxylon* sp.

All along in the history, droughts have been a major cause of human misery and land degradation. Major advancers in degradation of vegetation occur during drought period. Saxena (1993) provided graphic details of degradation of grasslands dominated by tussocky perennial grasses during the drought of 1980's thus "*These grasses are subjected to heavy grazing pressure. Continuous droughts rendered the clumps very weak due to exhaustion of stored food in rhizomatous parts and poor sprouting was observed from tussocks during the light shower that followed. Majority of tussocks failed to respond. The sprouted tussocks were grazed endlessly and majority of these got trampled and finally died. During third year no perennial stock of these grasses could be seen in grasslands. Low perennial grasses like Eleusine compressa and Dactylocteniu msindicum that appeared after recovery from drought, also did not perform well in the first two years and provided negligible above-ground biomass".* For people, drought means personal misery, death or distress sale of livestock, increased indebtedness and migration. However, improved economy of the State and the ability to provide relief has significantly softened the adverse socio-economic effect of drought situations.

## Severity and Consequences of the Problem of Land Degradation

As mentioned earlier, assessment of desertification has been an ongoing area of study world over. CAZRI has been making inferences based on manifestations observed under various land use and management system. Besides, actual measurements were made also with instruments. Because of its widespread nature, wind erosion received maximum attention and later studies made full use of remote sensing tool as well. However, since there is a lot of year to year variation in wind speed, the rate of change was non-uniform. An early estimate (CAZRI, 1976) put the area affected by wind erosion at 0.14 million square kilometer or ~70 per cent of the geographic area of Thar, a breakup of which showed 14 per cent as severely affected, 28 per cent moderately so and 27 per cent as slightly affected. Kar *et al.* (2009) with intensive use of remote sensing outputs has further refined this information for early 2000's.

Several experts have indeed considered wind erosion as the most serious, partly as it was coming in way of the major canal irrigation project. But some have differed on the ground that despite its seriousness, there has not been any reduction in crop area because of lands going out of cultivation or decline in yield of crops on rainfed lands. Undoubtedly, an episode of an above normal wind regime does cause loss of land level but the same is largely within the capacity of individual farmers to remedy with tractor. Likewise crop yield loss is temporary as the affected lands have the capacity to recover. Moreover, the past few decades have seen an unexpected decline in wind speeds. Thus, though there is some loss due to wind erosion, fortunately it is not as serious as in the past.

The grazing lands are a different case and these have indeed undergone severe degradation. Such lands had got seriously depleted over the past 4-5decades and presently these continue just 10 to 35 per cent of their potential. In the western drier part, where the coverage of grazing lands is also much larger, the degradation process started later but today over 90 per cent of these lands are severely degraded also. Though, the problem is indeed serious and the same is manifest in the large decline of cow as per cent of livestock and an increase in that of goat, a state of crisis in animal husbandry has not happened. Goat is far more versatile and can freely browse on thorny and other shrubs not suitable for cattle and sheep. Besides, irrigated farms generate a lot of by-produce, which finds its way to the market for animal holdings of pastoralists and rainfed farmers.

Though waterlogging and soil salinization was becoming a menace in the precious canal irrigated areas in the past, control on water supplies, dewatering and change in cropping pattern has been able to contain the problem. However, dwindling of groundwater resources is indeed serious. Groundwater irrigated areas account for more than half of the total cropping under irrigation. Aquifers with acceptable quality of water are few and these are undergoing decline of water table by 1-3 meters annually. In some areas the aquifers are already exhausted but if the irrigated area has not suffered a setback, it is due to shifting to as yet under-exploited sites or by shifting to sprinkler method of irrigation. But this situation will sustain for another 5-10 years only, after which irrigated agriculture will start dwindling considerably.

## Desertification Control Efforts and their Analysis

Not necessarily from desertification point of view, the central government, concerned with the plight of farmers in climatically disadvantaged regions of the country, launched in the year 1974-75 a nationwide program " Drought Prone Area Program (DPAP)". Little later, in the year 1977-78, was started "Desert Development Program (DDP)" exclusively for desert region. For some time both these activities operated concurrently in Thar Desert, but from early 1980's DPAP funding was withdrawn but the DDP, funded mainly by Central Government, continues to function and over time the coverage was has been expanded to arid areas in the southern part and to cold desert in the north. Besides, the technical content of the activity has been broadened also. Some non-governmental organizations (NGOs) are undertaking natural resource regeneration activities as part of their rural development activity. In initial stages of the Program, development of surface water resources through construction of small-medium sized reservoirs and drinking water supply sectors were also components of this activity. Over time, main focus has been on desert afforestation, sand dune stabilization, pasture land improvement and soil and water conservation.

Afforestation was one of the major areas of activity and it covered the canal irrigated areas, where protective of vegetation was critical to the successful functioning of the infrastructure. Besides, there were several other related activities as listed in Table 2.1.

**Table 2.1: Physical Achievements of Afforestation Activity in Thar Desert**

| Activity | Area Covered '000 ha |
|---|---|
| IGNP canal-side afforesttaion | 136.0 |
| Road side plantation | 141.0 |
| Block plantation | 3.2 |
| Fuel wood plantation | 42.8 |
| Shelter belt plantation | 3.6 |
| Total | 326.6 |

Includes also afforestation activity undertaken by Eco-Task Force of Territorial Army.

A technology has been developed quite early for stabilisation of instable and shifting sand dunes. A king pin of this was the tree *"Acacia tortilis"*, which was hardy and yet fast growing.

As of now ~119 thousand ha area has been treated, much of it in IGNP command and along its water distribution system. However, this activity expanded to some extent only in other parts because of reservation from local people, who did not want exclusion of the land from their use.

Pasture development has been another major activity. A technology for rehabilitation based on fencing to prevent damage during establishment period, land preparation and reseeding with appropriate perennial grasses with an element of shrubs and trees already existed with CAZRI. The grasses are hardy, efficient utilizers of moisture in terms of biomass production and are highly palatable. The same applies to chosen shrubs and trees. The technology had also been demonstrated successfully for varied ecological settings in the Thar Desert. As of now this development has taken place over ~200 thousand ha. Besides, NGO's have contributed another ~10 thousand ha. Though this effort succeeded in improving the worst of grazing lands, the developed lands deteriorated soon thereafter because of poor control on grazing, which was critical for continuous good health of developed pastures.

## Desertification and Development

Right from the beginning of the independence period, the governments of the day had a major commitment to grow more food and undertake socio-economic development of rural masses. Universal literacy, drinking water security, public health mission, rural employment, skill development, poverty alleviation and rural housing have been major efforts. Rural infrastructure, communication and use of modern information technology received also a high priority. Urbanization, tourism development, industry, and growth of service sector have seen major developments. Irrigation development, besides increasing and stabilizing land productivity, created a large demand for labour, which has stood in good stead for rural poor and landless. These activities have increased both the rural and urban labour wages (Dhir, 2003). In fact wages in Thar Desert today are higher than those in other parts of Rajasthan and the region is attracting labour from outside the region. The stress due to drought has got greatly reduced due to the ability of the State to provide necessary relief.

Wind regime, for as yet unexplained reasons, has become milder and therefore the problem of sand drifts is not as grievous as it was 1970's and 1980s. But whether it is a temporary phenomenon or a durable one remains to be established. Meanwhile global warming has added a new dimension in destiny of environmentally marginal areas. Most studies conclude that rainfall may not decrease but it is certainly going to get more erratic and torrential in character. Further rise in potential evapo-transpiration needs of crops is going to increase their water need and thus cause an increased stress. Further rise in temperature, is going to reduce the length of growing season during winter season and thus adversely affect their yield potential. Some models suggest that wind regime is going to strengthen in near future and in that case wind erosion will become a menace once again.

## Summary

Summing up all the above, though the problem of desertification in its various manifestations is real and several action programs have also been undertaken by individual countries, socio-economic linkages, community organization and large dependence of people on land productivity have come in the way of rehabilitation efforts. Sustainability of the effort is constrained not just by natural limitations but also by the inequity, poverty and capacity of the societies who are directly impacted and whose actions are the major drivers of the problem. Though, biodiversity and efforts to reduce global warming have significant convergence, there are areas of conflict also. UNCCD in its White Paper has expressed the situation as "there is widespread consensus that the problem of desertification, land degradation and drought remain inadequately addressed in today's political agenda at the global levels" (Low, 2013).

In respect of Thar Desert, many interventions have been implemented such as irrigation development, particularly from water sourced outside the region, which have reduced dependence on rainfed farming and in the process also stabilized the landforms, which were once threatening the well-being of the people. Public food distribution system has also practically done away with possible starvation and deprivation. Similarly, there have been developments in rural infrastructure, growth of non-farm employment, public health, literacy and rural housing have helped in minimizing pressure on land in the region.

## References

1.  Anonymous (2010). Degraded and Wastelands of India: Status and Spatial Distribution. Indian Council of Agricultural Research New Delhi and National Academy of Agricultural Sciences New Delhi. 76 p.

2.  Bojö J. (1996). The costs of land degradation in Sub-Saharan Africa.*Ecological Economics*16: 161-173.

3.  Bradley D. and Grainger A. (2004). Social resilience as a controlling influence on desertification in Senegal. *Land Degradation and Development* 15: 451-470.

4.  Cheng L., Cui, X., Gong, L. and Lu, Qi.WP (2013). Methodologies for Valuating Desertification Costs in China- Case Study 1. In "Economic and Social impacts of desertification, land degradation and drought, White Paper I". UNCCD 2nd Scientific Conference. United Nations Convention to Combat Desertification (UNCCD) Ed. Low, P.S. Bonn, Germany: 57-60.

5.  Darkoh, M.B.K (2003). Desertification in the drylands: A review of the African situation. *Annals Arid Zone* 42: 289-308.

6.  DDC (2003). Land Tenure Reforms and the Drylands- The Global Drylands Imperative-Second Challenge Paper Series. United Nations Development Programme Dryland Development Centre, Nairobi.

7.  Dhir, R.P. (1995). Problem of desertification in arid zone of Rajasthan-A view. *Desertification Control Bulletin* 27: 45-52.

8. Dhir, R.P. (1999). Natural resources and desertification: Need for refinement of behavioral interpretations: In *"Management of Arid Ecosystem"* (Eds. A.S. Faroda, N.L. Joshi, S. Kathju and A. Kar), pp. 1-4. Arid Zone Research Association of India and Scientific Publishers, Jodhpur.

9. Dhir, R.P. (2003). Thar Desert in retrospect and prospect. *Proceedings Indian National Science Academy* 69 Part A: 167-184.

10. Dhir, R.P. (1982). The human factor in ecological history. In *Desertification and Development* (Eds. B. Spooner and H.S. Mann), pp. 311-332. Academic Press, London.

11. Dregne H. (1983). *Desertification of Arid Lands*. Harwood Academic Publishers. New York.

12. Dregne H., Kassas M., and Rozanov B. (1991). A new assessment of the world status of desertification. Desertification Control Bulletin 20: 6-18.

13. Eriksen, H.S. (2003). Biophysical processes and livelihood interactions in the drylands*Annals Arid Zone* 42: 231-254.

14. Faroda, A.S. (1998). Arid zone research- An over view. In *Fifty Years of Arid Zone Research in India* (Eds. A.S. Faroda, and M. Singh), pp. 1-16. Central Arid Zone Research Institute, Jodhpur.

15. Fleskens. L, Nainggolan, D. and Stringer, L. (2013). Economic assessment of DLDD in Spain- Case Study 2. In "Economic and Social impacts of desertification, land degradation and drought, White Paper I". UNCCD 2nd Scientific Conference. United Nations Convention to Combat Desertification (UNCCD) Ed. Low, P.S. Bonn, Germany: 61-62.

16. Fleskens, L., Irvine, B., Kirkby, M. and Nainggolan, D. (2012). Desertification Mitigation and Remediation of land (DESIRE) – A global approach for local solutions: Deliverable 5.3.1.Model.

17. outputs for each hotspot site to identify the likely environmental and social effects of proposed remediation strategies. 156 p.

18. Jiang, H.(2002). Culture, ecology and nqature's changing balance: Sandification on Mu Su sand land, Inner Mongolia, China. In "Global *Desertification: Do humans cause deserts* ?Eds J.F. Reynolds and M.S. Smith: 181-196 Dahlem University Press, Berlin.

19. Kar, A. Garg, B.K., Singh, M. P. and Kathju, S. (Eds. (2009). *Trends in Arid Zone Research in India*, 481p. Central Arid Zone Research Institute, Jodhpur.

20. Kar. A, Mohrana, P.C., Raina, P., Kumar, M., Soni, M.L., Santra, P., Ajai, Arya, A.S. and Dhinwa, P.S. (2009). Desertification and its control measures. In Trends in Arid Zone Research in India(Eds. A. Kar, B.K. Garg, M.P. Singh and S. Kathju), pp. 1-47. Central Arid Zone Research Institute, Jodhpur.

21. Low, P.S. (ed) (2013). Economic and Social impacts of desertification, land degradation and drought. White Paper I. UNCCD 2nd Scientific Conference, United Nations Convention to Combat Desertification (UNCCD). Bonn, Germany. 62p.

22. Middleton N.J. and Thomas D.S.G. (1997). *World Atlas of Desertification.*2nd Edn. Arnold, London.

23. Mortimore, M. (2003). Is there a new paradigm of dryland development *Annals Arid Zone* 42: 459-482.

24. Reddy, V.Ratna (2003). Land degradation in India: costs and determinants. *Economic and Political Weekly.* XXXVIII (44): 4700-4713.

25. Rocheleau, D.E., Steinberg, P.E. and Benjamin, P.A. (1995). Environment, Development, Crisis and Crusade.Ukambam, Kenya. 1890-1900 *World Development* 23: 1037-1051.

26. Samra, J.S. and Sharma, P.D. (2005).Quality of soil resources in India. In "Souvenir International Conference on Soil, Water and Environment Quality-Issues and Priorities" Indian Society of Soil Science, New Delhi: 1-29.

27. Saxena, S.K. (1977). Desertification due to ecological changes in the vegetation of Indian desert. *Annals of Arid Zone* 16: 367-373.

28. Saxena, S.K. (1993). Impact of successive droughts on the natural flora of Indian Arid Zone. In *Desertification and Its Control in the Thar, Sahara and Sahel Regions* (Eds. A.K. Sen and A. Kar), pp. 159-168. Scientific Publishers, Jodhpur.

29. Shankarnarayan, K.A. (1988). Ecological degradation of Thar Desert and eco-regeneration.In *Desert Ecology* (Ed. I. Prakash), pp. 1-14.Scientific Publishers, Jodhpur.

30. Spooner, B. (1982). Rethinking Desertification: the social dimension. In "Desertification and Development" Academic Press, London: 1-25.

31. Singh, S (1982). Types and formation of sand dunes in the Rajasthan Desert In "Oerspectives in Geomorphology (Ed. H.S. Shrma) 4: 165-183, Concept Publishers, Jaipur.

32. Smith, M.S. and Reynolds, J.F. (2003). Interactive role of human and environmental dimensions in their desertification debate *Annals Arid Zone* 42: 255-270.

33. Stringer L.C., Dougill A.J., Thomas A.D., Spracklen D.V., Chesterman S., IfejikaSperanza C., Rueff H., Riddell M., Williams M., Beedy T., Abson D.J., Klintenberg P., Syampungani S., Powell P., Palmer A.R., Seely M.K., Mkwambisi D.D., Falcao M., Sitoe A., Ross S., Kopolo G. (2012). Challenges and opportunities in linking carbon sequestration, livelihoods and ecosystem service provision in drylands, *Environmental Science and Policy* 19-20: 121-135. doi: 10.1016/j.envsci.2012.02.004.

34. Suresh Kumar, (1997). Vegetation of the Indian Arid Ecosytem. In *Desertification Control* (Eds. S. Singh and A. Kar), pp. 71-79. Agro Botanical Publishers (India), Bikaner.

35. Warren, A. and Olsson, L. (2003). Desertification: loss of credibility despite evidence. Annals Arid Zone 42: 271-289.

*Note*: Data on human population in different parts of the country over the census years and on land use cited in this paper are freely available from Census Reports of Government of India and from Directorate of Economics and Statistics, Department of Planning, Government of Rajsathan, Jaipur,)

## Chapter 3

# Water Management and Infrastructure Planning to Cope with Water Scarcity in Drought Prone Areas in India

*Geeta*

*Research Associate,*
*Centre for Science and Technology of the Non-Aligned and*
*Other Developing Countries (NAM S&T Centre),*
*New Delhi, India*
*E-mail:geetageet01@gmail.com*

## Abstract

In drought prone areas, coupled with reckless abuse and increasing water demand, and due to growing population and unsustainable lifestyle, many countries are facing severe water crisis. Many developing countries including India will have to face crisis of food and water security in the near future. Certain parts of India are reeling under a severe drought condition and it is estimated that 30 per cent of the total geographical area of the country is undergoing degradation and drought. Categorized as one of the worst droughts in the history of India, drought in 2015-16 has affected more than 330 million people in more than 2.50,000 villages of 266 districts from 11 states in the country. It has had a devastating impact on people's lives as it affected water availability, livelihoods, food security, and natural resources and also put a huge burden on the exchequer. The impact is also visible in agriculture and the food grain production. The food grain production in 2015-16 decreased to 253.16 million tones compared to the average production in the preceding 5 years (2010-11 and 2014-15) which was 255.59 million tones. Nearly 15 states in India may face severe shortage of groundwater if we continue to exploit it continuously. In extremely hot summer and due to excessive pumping of groundwater, the water table is decreasing day by day by 6m or more each year. Every year poor farmers are committing suicide as their crops suffer heavily due to drought and insufficient water availability. State surveys revealed that the scale of damage done by drought was huge and the proportions of farmers who reported crop loss ranged from 60 to 94 per cent.

Application of innovative technologies, development of sustainable infrastructure and adoption of appropriate management strategies are required to obviate the problems of water scarcity in the drought affected regions. This paper deals with the problems of water scarcity, aridity, drought and desertification as well as water management strategies required to cope with the problems in water stressed areas and then identifies the need of emerging technologies and infrastructure for water management in such areas.

*Keywords: Drought, Groundwater, Water scarcity, Infrastructure, Planning, Water management.*

## Introduction

*Drought is a prolonged period of inadequate precipitation resulting in large scale damage to crops, resulting in loss of yield.* As per the guidelines of National Commission on Agriculture in India drought can be classified in three categories: Meteorological, Agricultural and Hydrological. Meteorological drought is defined as a situation when there is significant decrease from normal precipitation over an area (*i.e.* more than 10 per cent). Hydrological drought results from prolonged meteorological drought resulting in depletion of surface and sub-surface water resources. Agricultural drought is a situation when soil moisture and rainfall are inadequate to support healthy crop growth. In India drought take place in areas with high as well as regions with low rainfall. Drought is no longer meager scarcity or the absence of rainfall, but related to inefficient water resource management, water scarcity problems occur in Himalayan region also. Requirement of over 80-90 per cent of drinking water and over 50 per cent for irrigation is met from groundwater. In absence of effective and large scale rainwater harvesting only little recharge is taking place. Further, arid zone may be defined as the zones undergo chronic water deficit. However, there are larger areas in India with water shortage at one or another. The monsoon regime leaves the major part of the country dry for a period varying from 2 to II months on an average, but the inter-yearly variations are quite important. Whereas, the normal dry season, when rains are not expected, goes unnoticed, the extension of the dry spell beyond the time of onset of the annual monsoon season poses a severe threat to water management and agricultural operations. This is also the case when couples of weeks pass without receiving any precipitation during the rainy season. India is a member of United Nations Convention to Combat Desertification (UNCCD), and the Government of India in fulfillment of one of the obligations of the parties to the convention, submitted a National Action Programme to Combat Desertification to the Secretariat of the UNCCD in 2001. In India total area under desertification is 81.45 mha. Further, Water erosion (26.21 mha) is the most pronounced process, followed by aeolian processes (17.77 mha) and vegetal degradation (17.63 mha). Nearly one third of the country's land area (32.07 per cent) is undergoing processes of land degradation. Total area under land degradation is 105.48 mha.

## Water Availability and Requirement in India

As per the World Bank estimates, India's population is 1.27 billion and assumed to reach at 1.6 billion by the year 2050. Although food grain production has increased, in order to meet the ever growing demand by this population increase there is a need to increase the production of food grain. However land and water resources are limited, there is requirement an improvement in the productivity of crops. Alone agriculture sector accounts for approximately 70 per cent of the global freshwater withdrawal and approximately 90 per cent of its consumptive use.

On an average annual precipitation of India is about $4000km^3$. However Water availability in India differs considerably covering the regions, and over time. It is estimated that out of the 4000 km$^3$ water, 1869 km$^3$is average potential flow in rivers available as water resource. Out of this total available water resource, only 1123 km$^3$ is utilizable (690 km$^3$ from surface water resources and 433 km$^3$ from groundwater resources).

| | |
|---|---|
| Area of Country as per cent of World Area | 2.4 per cent |
| Population as per cent of World Population | 17.1 per cent |
| Water as per cent of World Water | 4 per cent |
| Rank in Per Capita Availability | 132 |
| Rank in Water Quality | 122 |
| Average Annual Rainfall | 1160mm (world average1110) |
| Range of Distribution | 150-11690 |
| Range Rainy Days | 5-150 days, Mostly during 15 days in 100hrs |
| Range PET | 1500-3500 mm |
| Per Capita Water Availability (2011) | 1545 m$^3$ |

*Source*: Water Resources at a Glance 2011 Report, CWC, New Delhi, (http://www.cwc.nic.in).

According to the international norms, when water availability is less than 1700 m$^3$ per capita per year country classified as 'water stressed', whereas water availability is less than 1000 m$^3$ per capita per year it is classified as 'water scarce'. The availability of surface water in India in years 1991 and 2001 were 2309 m$^3$ and 1902 m$^3$. Though, it has been estimated that per capita surface water availability is probably to be reduced to 1401 m$^3$ and 1191 m$^3$ against 5200 m$^3$ of the year 1951 in the country. Many parts of the country are reeling under drought and water crises; the situation is getting poor by the day. In India, per capita water availability has come down 70 per cent from 1951 to 2011 in period of 60 years.

It has been assessed by the 'Standing Sub-Committee for Assessment of Availability and Requirement of Water for Diverse Uses in the Country' total water requirement for different uses like agriculture, industrial and domestic to be about 813 BCM, 1093 BCM, and 1447 BCM by the year 2010, 2025 and 2050 respectively. However, the assessment of National Commission for Integrated Water Resources Development is the water requirement by the year 2010, 2025 and 2050 will be about 710 BCM, 843 BCM and 1180 BCM respectively and the higher demand would for Irrigation. As per the estimate of UN FAO, in 2010, the withdrawal of water in India by irrigation and livestock is 91 per cent. India has 4 per cent of usable water resources whereas it has 18 per cent of the world's population and is expected to face the water scarcity crisis.

# Challenges of Water Scarcity in Indian Agriculture

## Agriculture Sector is the Highest Consumptive Sector of Freshwater

Central Groundwater Board estimates show that in India 15 states may face serious shortage of groundwater with the continuity to utilize blindly. Presently it is difficult to get freshwater because of increasing pollution and domestic consumption, industrial use and irrigation; consequently there is a clash among these. Agriculture sector is the prominent in all economic sectors where water scarcity is higher. In India according to United Nations Food and Agriculture Organization (UNFAO, 2010), irrigation and livestock accounted for 91 per cent of water withdrawal which is higher than the global average. Most of water withdrawal came from groundwater, groundwater table is depleting very fast and in lack of quick recharge mechanism it is a very long drawn process. As per World Bank estimates in 2010, withdrawal of groundwater for irrigation is 60 per cent. According 10 state groundwater board officials, water tables are dropping by 6 m or more each year. It was learnt that farmers in the country, a generation ago, used bullocks to lift water from shallow wells in leather buckets. But now, they have to draw water from 300 metres below ground using electric pumps. The pumps powered by heavily subsidized electricity are working day and night to irrigate fields of more water consuming crops like rice, sugarcane and banana. This massive unregulated expansion of pumps and wells is threatening to suck India dry. Hence, managing water resources should attract due attention.

## Weather Irregularities, High Dependence on Rain Fed Agriculture and Polluted Water Bodies

Late monsoon early monsoon, prolonged arid spells and monsoon with small showers create drought and water logging problems. In India the distribution of rainfall across various regions differs with highest rainfall in Mawsynaran and lowest in Western Rajasthan. In lack of proper management of water resources the areas with highest rainfall also experience water scarcity. Therefore, over the years, water rich regions have become water scare and water stresses areas face water famines. Further, 67.5 per cent of the cultivated area is occupied by rain fed areas and it contributes 44 per cent of food grains and support 40 per cent of population and two thirds of livestock. Therefore, half of the cultivated area will continually depend on rain. Not only high dependence on agriculture is the challenge but also water bodies including rivers, canals, and wetlands are sorely polluted because of sewage and sludge, and waste products from industrial units.

Present in sequence the various water management techniques identified by the study (micro-irrigation, watershed management, rain water harvesting, traditional methods, urban methods *etc.* and relate to your area of study).

## Emerging Technologies and Infrastructure for Water Management Programme in India

Recognizing the importance of emerging technologies and infrastructure, the Indian government has taken various initiatives since 1992. In 2006, the government launched a Centrally Sponsored Scheme for micro-irrigation and after that upgraded to the National Mission on Micro-irrigation and was implemented through the year 2013-14. In India three area-based watershed Programmes for development of wastelands/degraded lands namely Drought Prone Areas Programmes (DPAP) was launched in 1973-74 to tackle the special problems faced by those areas that are constantly afflicted by drought conditions. Presently, 972 blocks of 195 districts in 16 States are covered under the Programme, Desert Development Programme (DDP) was launched in 1977-78 to mitigate the adverse effects of desertification. Presently, 235 blocks of 40 districts in 7 States are covered under the Programme and Integrated Wastelands Development Programme (IWDP) has been under implementation since 1989-90. The projects under the IWDP are generally sanctioned in areas that are not covered under DDP or DPAP.

## Micro-irrigation

Micro-irrigation is the application of small quantities of water directly above and below the soil surface; usually as discrete drops, continuous drops or tiny streams through emitters placed along water delivery line. With the usage of micro-irrigation systems conveyance loss, evaporation, runoff and deep percolation can be reduced. Proximity and focused application leads to higher water usage efficiency in micro-irrigation and also increase the area under irrigation as well as more usage of degraded land.

In India, after the adoption of micro-irrigation system increase in the area under irrigation was 8.41 per cent across 13 states. In the same survey, the sampled farmers indicated that 845.50 hectares of waste/degraded land was not being used for cultivation. However, after the adoption of the micro-irrigation system, the farmers were able to bring 519.43 hectares of such land under cultivation. In India electricity consumption in agriculture sector accounts for 20-25 per cent. In 2013-14 the consumption was 166,712mn kWh and government provided INR 66,988 crore (US$ 10.98 bn) in subsidies. In the year of 2013-2014, subsidies on fertilizers to the agriculture sector has been increasing, it accounts for at INR 71,251crore (US$ 11.68 bn). Application of Micro-irrigation system and water can improve fertilizers uses efficiency by 28.5 per cent on an average.

### Advantages of Micro-irrigation Systems

★ Micro-irrigation systems provide a high degree of water application uniformity. Due to the lower water requirement as a result smaller power units required significantly electricity savings have been estimated.

★ Direct use of fertilizers to the root increases the efficiency of usage. Runoff is minimized because of the low application rates, after the sufficient use of water deep percolation losses can also be minimized.

★ Due to controlled and targeted application of water, soil moisture can be maintained and this increases the crops productivity. Various surveys show that micro-irrigation systems helped increase the yield of fruits as well as vegetables and this in turns helps farmers income.

★ Application of micro-irrigation systems improves the efficiency of water use, helps in the introduction of new crops and overall irrigation cost brought down.

## Components of Micro-irrigation Systems

The different types of Micro-irrigation systems are all made up of the same basic components. Below is a short overview of the components of micro-irrigation systems.

### Water Source

Two type of water source exist for micro-irrigation these water sources are: groundwater and surface water. Many existing and potential water supply sources for irrigation systems are derived from surface water, which does not tend to have high levels of salts (with the exception of some coastal areas), and thus systems are usually less prone to formation of precipitates in drippers when using a surface water source. Surface water, however, tends to introduce biological hazards. If wastewater is being considered as a source, quality and clogging potential will vary depending upon the extent of treatment. Groundwater is generally of higher quality than surface water.

### Pumping Station

The selection of pump and motor is important aspect to deliver the correct pressure and flow rate at the highest possible efficiency.

### Flow Meters

Flow meter is the most important part of micro-irrigation system. Flow meter is used to determine the amount of water to apply, which, sequentially, is critical to efficient irrigation and scheduling. A propeller meter, which displays either the flow rate and/or total water applied, gives an accurate measurement.

### Filters

Filters are used to prevent the problems of clogging the drippers and drip tape holes resulting in obstructing the water to the plants by foreign particles, leaves, algae coming with water. Different types of filters are being used in drip irrigation systems like sand filters, screen filters, hydro cyclone filters and disc filters.

### Valves

These Valves are used to control the mechanisms of system; various types of valves are common: control valves, vacuum relief valves and check valves. To maintain a stable operating pressure in the system pressure-regulating valves are important.

### Fertigation Equipments

**Fertigation** is the injection of fertilizers, soil amendments, and other water-soluble products into an irrigation system. **Fertigation** are related to chemigation, the injection of chemicals into an irrigation system. Venturi, Injector and fertilizers tank are main component of fertigation. The **Venturi** effect is the reduction in fluid pressure that results when a fluid flows through a constricted section (or choke) of a pipe. Fertilizers tanks are used for fertilizer and chemical injection through drip and sprinkler irrigation systems. In this system part of irrigation water is diverted from the mainline through a fertilizer tank in a fluid or soluble solid form.

### Control Head

The **control head** contains valves to control the discharge and pressure in the entire system. It has filters to clear the water and these slowly add a measured dose of fertilizer into the water during irrigation. This is one of the important advantages of drip irrigation.

### Main Lines and Submains

Mainlines and submains are used to supply water from the control head in to the field. These made of PVC or polyethylene hose and lay down to ground because they easily degraded from direct solar radiation.

### Lateral Lines

These are source of drip laterals and are mad of LLDPE which quite flexible and strong. These convey water from sub main lines to root zone by dripper.

### Emitters or Drippers

Emitters or drippers are used to control the discharge of water from the lateral lines to the plants. These are spaced more than 1 meter with one or more emitters used for single plant. For row crops more closely spaced emitters may be used to wet a strip of soil

### Micro Sprinklers

Micro Sprinklers are used for irrigation of seasonal crops like vegetables, onions, potato, nurseries *etc.* These are made of engineering plastic and available in different connection options. It can be used to increase height in following products: multipurpose stake with adaptor for mini sprinkler, stake for mini sprinkler, mini sprinkler base. Having Sharp and self punching inlet it provides stable base for the installation of mini sprinkler.

## Watershed Management in India

Watershed is an area or ridge of land that separates waters flowing to different rivers, basins, or seas. This is the form of physical, hydrological and human resources. Watershed management involves the sensible utilization of natural resources for best creation with least damage to natural and human resources. Thus, Watershed management is a term used to describe the process of implementing land use practices and water management practices to protect and improve the quality of the water and other natural resources within a watershed by managing the use of those land and water resources in a comprehensive manner. Approximate 68 per cent of India's cropped area is under Rain-fed agriculture which provides livelihood to 480 million people. Hence watershed management is the only feasible choice to tackle production requirement in primary sector especially in arid and semi-arid climatic region.

In India, watershed development has been popular method to harvest, store and manage rainfall, runoff and stream flows at community level. Indian government adopted micro-watersheds programs on the base of traditional water management approaches since late 1980s. Then the guidelines for Watershed Development Projects became operational in 1995, and Planning Commission and National Rainfed Area Authority (NRAA) framed Common Guidelines, 2008 for watershed programmes for all Ministries/Departments based on the Parthasarathy Committee Report, other Committee's observations and past experiences. The Department of Land Resources, Ministry of Rural Development is implementing the Integrated Watershed Development Programme (IWMP) from 2009-10 with an objective to cover 55 million hectare of rain fed land by 2027. The program is being implemented in all the states of the country. In watershed development programs, rainfall and runoff harvesting schemes involve rehabilitating, construct small check dams and tanks, and groundwater recharge structure.

## Classification of Watershed Management

Classification of watershed Management depends upon the size, drainage, shape and land use pattern.

* Macro watershed (> 50,000 Hect)
* Sub-watershed (10,000 to 50,000 Hect)
* Milli-watershed (1000 to10000 Hect)
* Micro watershed (100 to 1000 Hect)
* Mini watershed (1-100 Hect)

## Objective of Watershed Management

* To increase the productivity of food, fodder and fuel.
* To conserve the natural resources, soil, water and land use.
* To store water, flood control and checking sedimentation.
* Sustainable development of eco-system.
* Improve infrastructure facilities.
* Improve the socio-economic status of farmers, cut down the risk and increase total income.

## Parameters of Watershed Management

The parameters of watershed management are:

* **Size:** It helps in computing parameters like precipitation received, retained, drained off.
* **Shape:** Different shapes based on morphological parameters like geology and structure which decides watershed characteristic.
* **Phisiography:** Consideration of land altitude and physical disposition.

* **Slope:** It controls the rainfall distribution and movement which is important for other factors.

* **Climate:** It decides the quantitative approach for watershed.

* **Drainage:** It determines the flow characteristics and so the erosion behavior which helped in water management.

* **Vegetation:** Information of species gives a sure ground for selection plants and crops for watershed management.

* **Geology and Soil:** Their nature determines size, shape, physiographic, drainage and groundwater conditions.

* **Hydrology:** It helps in quantification of water available.

* **Socioeconomic:** Statistics on people and socio-economic status are important in managing water.

## Component of Watershed

### a. Soil and Water Conservation Measures

The main objective of the soil and water conservation measures to maintain the quality of soil and allowing more runoff water to be soaking up held in soil profile and obstruct a long slope into different short ones, in order to maintain less than a critical velocity for the runoff water. Soil and Water Conservation Measures include contour bunding, terracing, and vegetative waterways.

### *Contour Bunding*

Contour bunding is the farming practice of plowing and/or planting across a slope following its elevation contour lines. This is the most widely practiced soil and water conservation measure for light soil and low rainfall area. Contour bunding helps to reduce the length of slope for stopping soil erosion and seize water and allow more filtration of runoff to increase soil moisture. It has been reported that in western part of India, contour bunds have helped to save about 25 to 162 tons/ha of soil from erosion. In addition contour bunds have helped to increase soil moisture and recharge groundwater storage (Das, 2009).

### *Graded Bunding*

Graded bunds are the bunds or terraces laid along a pre-determined longitudinal grade very near the contour but not exactly along contour. Graded bunding is suitable for rain fed areas where rainwater is not absorbed due to high precipitation or low penetration rate of water in soil. Further, graded bund system helps to dispose of excess runoff safely from agricultural field.

### *Bench Terracing*

Bench terracing constructed on hilly areas at the slope gradient of in the range of 15-30 per cent. Bench terracing alters a steep land surface in to a flatten steps over the slop of the land. These terraces transform the erodible sloping lands into cultivated land and the extent of these terraces depends on gradient of slop, depth profile of soil.

### Grassed Waterways

Grassed waterways are associated with channel terraces for safe disposal of concentrated run-off, thereby protecting the land against rills and gullies. A waterway is constructed according to a proper design depending upon the discharge, slope, and soil type *etc.* These waterways are used for disposal of water collected by road ditches or discharge through culverts and for the prevention of gullies formation.

### Strip Cropping

Strip cropping involves cultivating a field into long, narrow strips which are alternated in a crop rotation system. It is used in steep sloppy areas when there is no alternative method of reduce soil erosion. The most common crop choices for strip cropping are closely sown crops such as hay, wheat, or other forages which are alternated with strips of row crops, such as corn, soybeans, cotton, or sugar beets. The forages serve primarily as cover crops. In certain systems, strips in particularly eroded areas are used to grow permanent protective vegetation; in most systems, however, all strips are alternated on an annual basis.

### b. Erosion Control Measures for Non-Agricultural Land

Soil is crucial for many aspect of human life so soil conservation is the prevention of soil loss from erosion or low fertility caused by over usage, acidification, salinization or other chemical soil contamination. Various types of methods can be used for the soil erosion control for non-agriculture land like;

### Contoured and Staggered Trenches

Contour trenching is a technique that applied in arid areas for water and soil conservation and high agricultural production. Artificially trenches dug along the contour lines then water move towards the lower position of the hills is perpetuated by the trench and is gain access to the soil below. Main advantages of this technique are; rain water does not immediately run off the hill, crops do not suffer from water shortage because water balance enhanced and fertility of soil is not lost by water and wind erosion. The shape and size of trenches depend on slope of hills basically adopted for hill slope >20, however manually dug trenches are smaller than trenchers dug by machine.

### Gully Control Structure

Gully erosion develops into deeper cracks or ravines in extreme cases that remove the soil along drainage lines by surface water runoff. After started gully erosion it will continue to move by head ward erosion therefore temporary check dams are usually made to control gully erosion. These check dams may be permanent, semi-permanent or temporary. The structure of temporary check dams are made up of brush wood, wire and poles. Semi-permanent check dams made up of gabion structure and permanent check dams are concrete.

### Gradonies

Gradonies are steeply inward sloping narrow bench terraces constructed on contours. Generally, gradonies are appropriate for afforestation in uniformly steep sloping lands.

### c. Rainwater Harvesting and Water Management

In India, huge amount of water wasted during the monsoon every year just dries up or flow up. Rainwater harvesting is focus area for efficient management of the available water resources. Even if 10 per cent of annual water harvested properly, that will create a considerable amount of water to the tune of 1800 million liters.

### Methods of Rainwater Warvesting

1. **Rural Methods-** in rural methods generally traditional methods of rainwater harvesting are used and main objective of these traditional methods to facilitate irrigation for agriculture and manage water for domestic and drinking purposes. Presently various types of traditional structures like Tankas, Nadis, Bavdis, Raoats, Kuis, Virdas, Kunds, Khadins talabs and Johads are used to recharge groundwater levels.

2. **Urban Methods-** in urban methods more modernized system of rainwater harvesting is used. Various types of structures like roof catchment, gutters, down pipes, filter unit, storage tank and collection pits are used to conserve and manage the water.

Rainwater harvesting is a better solution of water problem in adequate water resources areas, it improves soil quality, increase soil moisture, and reduces the risk of flood. Rainwater harvesting reduces quantity of salt in the soil which is harmful to root growth of plants and this is helpful in root growth and increasing the drought tolerance of plants. This technology is not so much costly because construction, operation and maintenance are not very labor intensive.

## Conclusions and Recommendations

Sustainable water management is crucial for livelihood and of the people therefore there is need of a balance between economic and environmental consideration in watershed management. Water should be rational utilized and its careless wastage is to be removed. Agriculture sector is facing the problems droughts and floods all over India; it becomes significant to apply the emerging technologies in agriculture sector to solve the problems

## References

1.  Srivastava R. K., Sharma H. C., Raina A.K. (2010). Suitability of soil and water conservation measures for watershed management using geographical information system. *Journal of Soil water Conservation*, Vol. 9, No. 3, pp 148-153.

2.  Das Ghanshyam (2009). Hydrology and Soil Conservation Engineering. Kalyani Publication, New Delhi.

3. Lawrence J. Schwankl and Terry L. Prichard (2012). Irrigation Systems. Prune production Manual. Agriculture and Natural Resources Publication 3507. University of California, USA, pp. 100-107.

4. Sivanappan (1997). Researchgate.net report; National Mission on Micro-irrigation Impact study prepared for the Government of India, June 2014.

5. India Agri Stat; Mission for Integrated Development of Horticulture (MIDH); Department of Agriculture and Cooperation, Ministry of Agriculture, Government of India; National Mission on Micro-irrigation Impact study prepared for the Government of India, June 2014, retrieved August 31, 2015; Grant Thornton Analysis.

6. Drip Irrigation Handbook (2015). Understanding the Basics Version 001.02 - 2015 January 2017, Israel https: //www.netafim.com.au/Data/Uploads/Netafim_ Drip per cent 20Irrigation_Understand per cent 20the per cent 20Basics_Jan17 per cent 20 per cent 20v1-1 per cent 20LR.pdf (accessed 28 May. 2017).

# Chapter 4

# Facing Drought Problem in Indonesia

## Nora Herdiana Pandjaitan and Titiek Ujianti Karunia

*Civil and Environmental Engineering Department,*
*Bogor Agricultural University, Darmaga,*
*Bogor, Indonesia*
*E-mail: norahp@apps.ipb.ac.id*

## Abstract

While some areas in Indonesia are suffering from seasonal flood during the rainy season, other areas are hit by drought. Generally, floods occur spontaneously but drought comes slowly but insidiously.

One of the main impact of climate change is the shifting of drought intensity. This generates problems to the agriculture sector, specifically to paddy and palawija (crops planted as 2nd crops in dry season). Some researches in Java showed that the drought lasted for around nine month with rain intensity less than 220 m/month. For the past three decades, the drought intensity has changed greatly, especially on paddy fields. Drought for paddy during dry years are somehow more evenly distributed compared to drought for palawija.

Besides the impact of drought on agriculture, electricity production is adversely affected also due to poor water storage. This condition consequentlly causes the rotating electrical black-outs.

The objective of this study is to examine various interventions on mitigating the impact of drought in Indonesia. The study was carried out through analyzing existing drought condition in Indonesia and proposing several alternative solution. The method used in this research was a descriptive method by analyzing secondary data obtained from related institution and literatures.

One of the priority programs in Indonesian government's agenda is supporting irrigation to promote food security. According to this and facing the problem of water shortage the goverment had planned some programs such as improving water availability for irrigation, and improving the structure and infrastructure for irrigation.

The study recommends that some efforts to reduce the toll of drought are by increasing infiltration rate to get zero run off, adding reservoir/water tank facilities and increasing water use efficiency. Other feasible options are controlling land use change, scheduling sowing season, controlling groundwater use, reusing and recycling water, developing early drought warning systems as well as improving community's involvement in the program.

*Keywords: Drought, Mitigation, Food security, Agriculture.*

## Introduction

One of the most common disasters occurring in Indonesia is prolonged drought with high temperature, which is caused by the low temperature in Pacific Ocean and lower layer atmosphere called El-Nino. Indonesia is a tropical country with two seasons occurring alternately, dry season and rainy season. The state of both seasons is highly defined by meteorological condition. Imbalanced meteorological condition may cause varying natural disaster. Natural disaster is a combination of three elements; threat, susceptibility, and the ability to create the disaster. According to regulation number 24 in 2007 (BNPB, 2007),disaster is an event or chain of events which are threatening and disturbing the lives and livelihood of the surrounding community, either by natural factors and/or unnatural factors or man factors, which results in death tolls, environmental and material loss, as well as psychological damage.

In 2015 more than 1 million people, particularly in East Indonesia, were in need of food assistance as a drought-related food crisis affected most of the country. The rainy season did not start in December as expected, because after a short period of average rain in the first days of January, rainfall returned to well below-average in February.Crops were cultivated late, during a short period of average rainfall in January that had ended by the end of the month. If the rainy season does not take place, these late crops are expected to fail.

**Table 4.1: Indonesia' Current (2009) and Project of Water Budget (2015 and 2030)**

| No | Area | Supply (S) | Demand (D) | Balance 2009 (S − D) | Balance 2015s (S − D) | Balance 2030s (S − D) |
|----|------|-----------|-----------|---------------------|----------------------|----------------------|
| 1. | Sumatra | 111,077.65 | 37,805.55 | 73,272.10 | 48,420.07 | -67,101.34 |
| 2. | Java-Bali | 31,636.50 | 100,917.77 | -69,281.27 | -118,374.36 | -454,000.33 |
| 3. | Kalimantan | 140,005.55 | 11,982.78 | 128,022.77 | 118,423.17 | 88,821.14 |
| 4. | Sulawesi | 34,787.55 | 21,493.34 | 13,294.21 | 13,490.80 | -21,021.99 |
| 5. | Nusa Tenggara | 7,759.70 | 2,054.04 | 5,705.66 | -17,488.89 | -67,848.68 |
| 6. | Moluccas | 15,457.10 | 540.23 | 14,916.87 | 12,648.91 | 9,225.75 |
| 7. | Papua | 350,589.65 | 385.58 | 350,204.07 | 325,937.74 | 315,647.73 |

From data of Indonesia's current water budget and water budget projection on the table, Jawa-Bali dan Nusa Tenggara had a water balance problem since 2009 and it becomes worse in 2030 (Bappenas 2010). In 2030 Sumatera was predicted will have a bigger demand than supply. Figure 4.1 showed that water supply was predicted decrease in the future but on the contrary water demand increase.

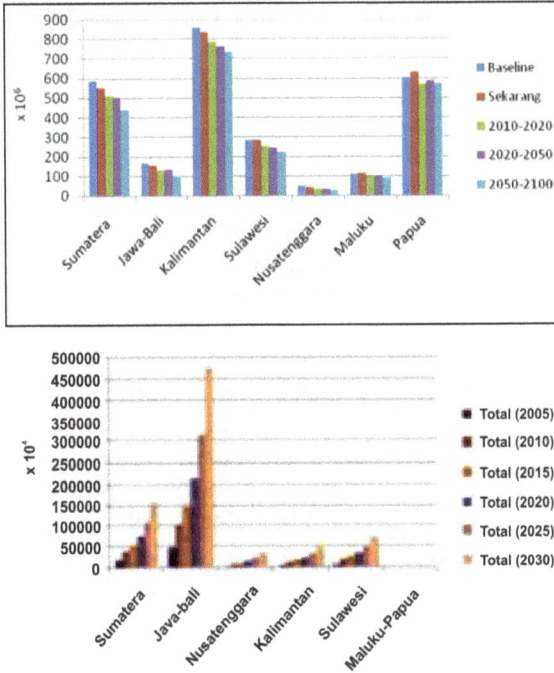

**Figure 4.1: Water Supply (Top) and Demand (Below) in Indonesia (Bappenas 2010).**

The method used in this research was a descriptive method by analyzing secondary data obtained from related institution and literatures.

## Drought Status in Java

The area of Java Island is 138,379 km², merely 7 per cent of the total area of Indonesia. However, the island holds a very important position as it is home to 59 per cent of Indonesia's population with a dcensity of 1,413 people/km². From economics perspective, Java has contributed 60 per cent to the GNP, which makes the island as the center of development in Indonesia. On the other side, the supporting potential of Java is steadily decreasing, as disasters and environmental damages become even more prominent. The main factors of natural disaster, apart from climate change and global warming, are the number and distribution of people, resources management, as well as regulations and government bodies regarding the management of these resources.

One of the parameters of drought is environmental change in crop field. Based on the data from Ministry of Public Works (KemenPU2007), drought in Java has reached 11.6 per cent of 1.8 million acres of the target area in the second planting period (MT II), which is around 209,332 acres. The biggest drought occurred in West Java, affecting 141,793 acre field or 22.5 per cent of MT II target, with 11,590 acres suffering from crop failure (puso). Drought in Central Java was quite wide, affecting 47,823 acres or 20.3 per cent from total MT II. In East Java, drought was relatively smaller, only 3.1 per cent of total MT or around 14,706 acres. In Banten, drought occurred for as big as 5000 acres in Pandeglang Regency.

In 2011 drought-susceptible index for all areas in Indonesia had been mapped (BNPB 2007). Based on that index West Java had score 19-24 and classified in the high category. In 2009 about 8,577 acres of paddy in West Java suffered from drought and 105 acres suffered from puso or crop failure.

The rainfall in Java is bigger than other areas as a consequence to global climate change (Pribadi and Sengara, 2010). Rainfall in dry season showed decreasing trend from 1 mm to 9 mm per season per year, whereas rainfall in rainy season showed decreasing trend from 1mm to 50 mm per season per year. Even in East Java there are some areeas with increasing trend from 1mm to 10 mm per season per year (Soetamto, 2009). It is projected that average rainfall in 2010-2020 will decrease compared to 1978-2007 and average rainfall in 2016-2020 will also be drier compared to 2010-2015. Danger of drought usually happens insidiously according to the drought characteristic. Community and surrounding environment responded to catastrophe slowly with the decreasing water reservoir.

In 2003 in areas outside Jabodetabek there are 77 per cent regencies suffering from water deficiency in 1 to 8 months within a year period (Bappenas 2005). It is estimated that by 2025, the number of regencies with water deficit will increase up to 78.4 per cent with the deficiency occurring between 1 to 12 months, or all-year deficiency.

Figure 4.2 showed level of drought in Java and Bali which was evaluated in one week, from 18 June to 25 June in 2015. Of those areas suffering from water

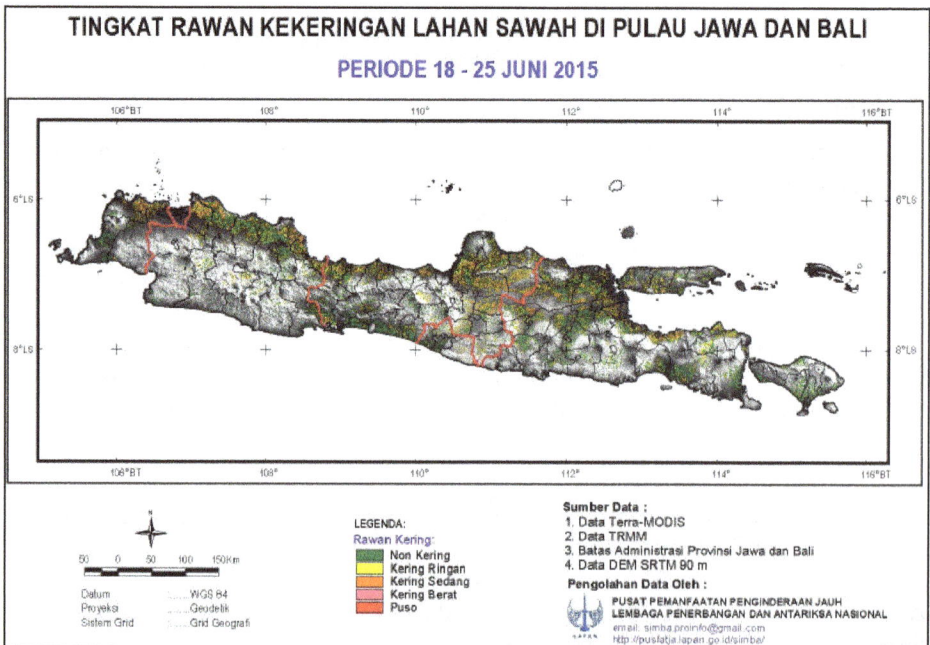

**Figure 4.2: Level of Drought-Susceptibility Index in Crop Field in Java and Bali.**

deficiency, 38 regencies/cities have water deficit for 6 months or more, classified as high-deficit. Moderate deficit is defined as water deficiency occurring for 4 to 6 months, whereas mild deficit occurring for up to 3 months. Normal condition is defined as not having any water deficiency within a year period.

Drought situation in Java is becoming even more concerning as the seasonal rainfall for Java is predicted to experience significant decrease in 2020-2080 (Boer *et al.*, 2007) and accompanied with increased intensity, coverage and severity of the drought (WMO 2009). El Nino/Southern Oscillation (ENSO) and Indian Ocean Dipole (IOD) positive phase are associated with drought in areas with monsoon climate, such as Java.

## Impact of Drought

The main factor of prolonged drought is water storage decreased due to climate. Other supporting factors that may intensify the drought are non availability of water reservoir to be used as buffer during dry season. Excessive and inefficient water exploitation will worsen the deficiency in surface water and soil water. Water absorbent areas, which are used as infrastructure development, will hasten the route of water to return to the sea. This consequently will reduce the water storage significantly. Increased water demand due to population development will also magnifies the scale of the problem especially in densely populated areas, such as Java.

According to Kelompok Kerja Air Minum dan Penyehatan Lingkungan Indonesia, water availability in Java is only 1,750 $m^3$/kapita/year in 2000, and will continue to decrease until 1,200 $m^3$/kapita/year in 2020. The minimum availability standard is 2,000 $m^3$/kapita/year.

Drought in Indonesia in the past two centuries has become the third biggest natural disaster after flood and fire (BNPB 2011). In 2015 almost all areas in Indonesia experience drought. Based on BMKG satellite imaging, no rain potential cloud was detected. In 2015, moderate drought occurred in 2 areas in Java.

Based on water resource in Java, it is showed that drought occurred in areas that can no longer contain the water demand pattern. According to Pawitan *et al.* (1996) long-term change in Citarum dam in 1896-1994 showed decreasing trend of rainfall of 10 mm/year and followed by decreased of runoff of 3mm/year. These changes happened due to the conversion of land use from forest into other function since twentieth century and the expansion of this use since the past three decades. The impact of this functional shift of forest in massive scale also changes the hydrological function of DAS. Hatmoko (2009) reported decreased water discharge in most of big rivers in Java, especially those in downstream.

## Drought Mitigation

There are solutions to reduce the impact of drought based on the time, whether short term, mid term, or long term. However there are usually not many options in addressing drought for short term (within one dry season). Available alternatives are usually increasing the usage of groundwater and artificial rainfall.

**Figure 4.3: Small Dam or Embung Using PVC Terpal.**

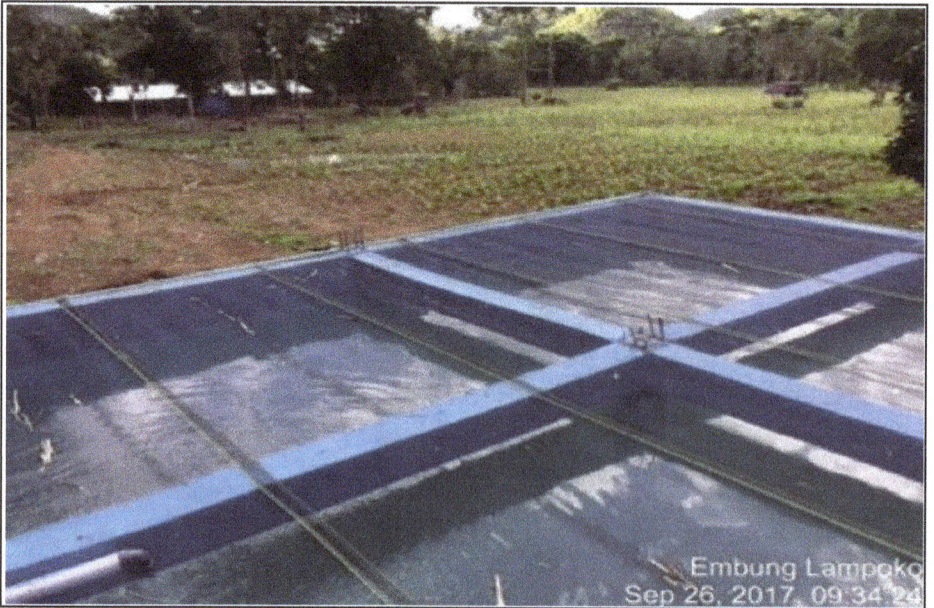

**Figure 4.4: Concrete Small Dam (Embung).**

**Figure 4.5: Long Storage.**

Mid term efforts may include revitalizing lakes, dam, small reservoir (embung), as water reservoir to store water when rainy season comes. Creating new embung in drought-susceptible areas may increase water storage, improve drainage to prevent and reduce leakage, and to prolong water cycle before returning to the sea by reusing and recycling water. In addition long storage can also be constructed on field as surface storage and shallow well as recharge well in residence area.

For long term measure, drought prevention can be done comprehensively and better-planned, for example by re-arranging special use with taking water cycle into consideration by doing reboisation and mass planting of trees in drought areas. Dam maintenance is also necessary so that when rainy season comes, it can contain enough water to be used during dry season. Other feasible options are controlling land use change, scheduling sowing season and developing early warning of further drought as well as improving community's involvement in the program.

New regulations from government are needed to improve better mitigation strategy and to reduce disaster, for example:

1. Arranging government rules regarding data delivery system from local to the center of data processing.
2. Creating local regulations to define scale of priority of water usage by taking into account historical right and justice.
3. Creating drought posts in local and national level
4. Creating special budget for development or maintenance of climate observation in drought areas.

## Conclusions

Indonesia has a serius drought problem specially in Java-Bali and Nusa Tenggara due to climate change and population development. To solve this problem, the government had built a lot of small reservoirs (embung) particularly for irrigation, conserved recharge area and small lakes (situ) and also empowered the community to make shallow well in residence area to store water during rainy season. Besides, indonesian people must learn to use less water and to do water reuse and recycle, to increase water efficiency

## References

1.  [Bappenas] Badan Perencanaan Nasional (2005). Studi prakarsa strategis SDA untuk mengatasi banjir dan kekeringan di Pulau Jawa. (not publish). Jakarta. Bappenas.

2.  [BNPB] Badan Nasional Penanggulangan Bencana (2007). Undang-undang Republik Indonesia Nomor 24 Tahun 2007 Tentang Penanggulangan Bencana. Jakarta. BNPB.

3.  [BNPB] Badan Nasional Penanggulangan Bencana (2011). Peta Indeks Risiko Bencana Kekeringan Untuk Seluruh Wilayah di Indonesia. Jakarta. BNPB.

4.  Boer, R., Sukardi, D. Hilman *et al.* (2007). Climate Variability and Climate Change And Their Implications In Indonesia. Jakarta.Ministry of Environment.

5.  [BPS] Badan Pusat Statistik (2005). Jumlah Kepadatan Penduduk Pulau Jawa. Jakarta. BPS.

6.  Hatmoko, W. (2009). Tren perubahan debit andalan sungai di Jawa. Makalah Lokakarya Tren Perubahan Curah Hujan dan Debit Sungai sungai di Indonesia, Program Penguatan IPTEK Adaptasi dan Mitigasi Perubahan Iklim KNRT, 21 April 2009.

7.  [KemenPU] Kementerian Pekerjaan Umum (2007). Bencana Kekeringan di Indonesia. Jakarta. Kementerian Pekerjaan Umum.

8.  Pawitan *et al.* (1996). Keseimbangan Air Hidrologi di Indonesia Menurut Kabupaten (*Hydrology Water Balance of Indonesia*).Bogor. FMIPA IPB.

9.  Pribadi K., Sengara W. (2010). Pengurangan Risiko Bencana, dalam buku Mengelola Risiko Bencana di Negara Maritim Indonesia. Bandung. Institut Teknologi Bandung.

10. Soetamto (2009). Perubahan Pola Musim dan Curah Hujan di Indonesia, dalam Laporan Identifikasi Dampak Perubahan Iklim Pada Sumberdaya Air. Jakarta. Kementerian Riset dan Teknologi dan IPB.

11. [WMO]World Meteorological Organization (2009). Experts Agree on a Universal Drought Index to Cope with Climate Risks. Press Release No. 872, 2009. www.wmo.int. Tanggal Akses : 25 Agustus 2014.

## Chapter 5

# Sulfuric Acid as a Chemical Amendment for Increasing Reclamation Process of Salt Affected Soils of Iraq

*Baydaa H.A. Al-Ameri, Suad A. Al-Saedi,*
*Ibrahim B. Razaq and Ayad Gh. Rasheed*

*Soil Chemistry and Desertification Department,*
*Office of Agricultural Research,*
*Ministry of Science and Technology, Baghdad, Iraq*
*E-mail: baydaa.2012@yahoo.com*

## Abstract

Diluted solutions of sulfuric acid were evaluated as a leaching solution in comparison with water for reclamation of two major types of salt affected soils of Iraq, namely the Shura and the Sabakh. Solutions of 0.05 N and 0.1 N sulfuric acid resulted in remarkable increase in the hydraulic conductivity (HC) of both soils over that when water was used as a leaching solution. Electrical conductivity of the leachate of both soils leached with water was less than that when these were leached with either 0.05 N or 0.1 N acid solutions. Also, total salts remained in soils leached with water were markedly higher than that leached with the acid. Amount of $Na^+$ removed as an average of both the soils was 62 per cent, 50 per cent and 42 per cent under 0.1N, 0.05N $H_2SO_4$ and distilled water respectively. Thus leaching of such soils by appropriate solution of sulfuric acid is highly efficient in removing sodium salts out of soil profile. And for that matter it is highly recommended for use in reclamation projects in Iraq.

*Keywords:* *Reclamation of salt affected soils, Sulfuric acid as an amendment, Leaching solution composition.*

## Introduction

Salt affected soils occupy about 70 per cent per cent of agricultural land of the Lower Mesopotamian Plain of Iraq (Al-Tai 1971). These soils are of saline and saline alkali type either with a white salt crust on the surface or a dark brown color with a high content of deliquescent salts (Buringh 1962). The first one is locally known as a Shura/Shure, while the second is known as a Sabakh soils. Intensive reclamation projects for these soils were initiated in the early seventies. Reclamation process involved lowering the depth of water table and applying enough volume of water on the surface to remove the excess salts. These efforts, however, encountered real problems like slow reclamation process, re-salinization of the reclaimed area and the considerable loss of land productivity.

Low efficiency of reclamation process could be attributed to leaching with a high quality water or with a water of low electrolyte concentration which had been reported to induce a considerable decrease in soil permeability and consequently in hydraulic conductivity (HC) (Quirk and Schofield, 1955; McNeal *et al.*, 1968; Shainberg *et al.*, 1981; Shainberg *et al.*, 1989). Resalinization of reclaimed soils may arise from inefficient removal of exchangeable sodium (Miyamoto *et al.*, 1975, Kirda *et al.*, 1974; Papadopoulos 1985). Therefore, this work was designed to investigate the possibility of increasing the efficiency of reclamation process by using an appropriate concentration of sulfuric acid to liberate Ca through reaction with native calcium carbonate to the leaching solution.

## Materials and Methods

The soils studied were clayey alluvial salt affected of Salman Project, located 20 Km south east of Baghdad, Iraq. Soils were sampled from 0-30 and 30-60 cm depths and their analytical data are shown in Table 5.1. The homogenized soil of each depth was packed in plexiglass columns of 70 cm length and 12 cm inner diameter. The bulk density of the soil in the columns was $1.48 + 0.05$ g/cm$^3$. Six columns were prepared from each soil;. two columns of each soil were assigned to one of the following leaching solutions; distilled water (DW) and 0.05 N H$_2$SO$_4$ and 0.1 N H$_2$SO$_4$ solutions. Continuous ponding of 5 cm depth of either leaching solution on the surface of the soil for the entire leaching period was maintained as a leaching method. Effluent was collected daily for which volume, EC, pH, Na, Ca and Mg, were determined. The hydraulic conductivity of the soil was also determined throughout the leaching period. All measurements were conducted according to Page (1982).

## Results and Discussion

Solutions of 0.05 N and 0.1 N H$_2$SO$_4$ (Figure 5.1) had a remarkable effect in maintaining HC of both soils above that of water. The HC of Shura soil when 1 pore volume of either water, 0.05 N, or 0.1 N H$_2$SO$_4$ passed was 0.078, 0.082 and 0.09 cm/hr respectively. Hydraulic conductivity of Sabakh soil was less than that of Shura soil. An exponential decrease in hydraulic conductivity was observed with the amount of water passed through. Hydraulic conductivity of Sabakh soil when either 0.05 or 0.1 N solution used was markedly higher than that when leaching solution was

## Table 5.1: Properties of the Soils Used in the Study

| Property | Shure Soil | | Sabakh Soil | |
|---|---|---|---|---|
| | Depth (cm) | | | |
| | 0-30 | 30- 60 | 0-30 | 30-60 |
| EC 1:1 dSm$^{-1}$ | 38.4 | 19.56 | 62.48 | 18.0 |
| pH 1:1 | 6.58 | 7.04 | 6.35 | 7.06 |
| CaCO$_3$ gKg$^{-1}$ | 302.2 | 311.4 | 308.46 | 317.46 |
| Gypsum gKg$^{-1}$ | 0.54 | 0.36 | 0.41 | 0.28 |
| CEC cmolKg$^{-1}$ | 21.8 | 24.65 | 24.6 | 23.28 |
| Sand per cent | 11.00.1 | 89.6 | 249.6 | 89.6 |
| Silt per cent | 28.0 | 300 | 280 | 320 |
| Clay per cent | 61.0 | 610 | 470.4 | 590.4 |
| Texture | Clay | Clay | Clay | Clay |
| Na mmolL$^{-1}$ | 268.12 | 154.77 | 333.07 | 129.25 |
| Ca mmolL$^{-1}$ | 29.45 | 7.69 | 117.42 | 19.12 |
| Mg mmolL$^{-1}$ | 88.53 | 33.81 | 166.78 | 33.68 |

distilled water. The greatest the salt content of the leaching solution showed the greatest hydraulic conductivity of soil. This is in agreement with Shainberg (1981) who reported that high electrolyte content of soils during the reclamation was very important for maintaining high hydraulic conductivity.

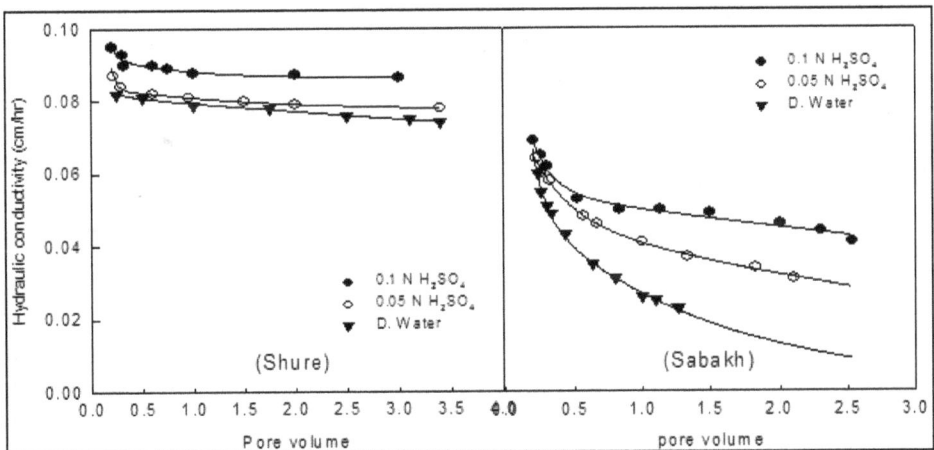

Figure 5.1: Effect of Leaching Process on the Hydraulic Conductivity of Soils Specified.

Efficiency of H$_2$SO$_4$ in reclamation process of Shura and Sabakh soils of Iraq was also evaluated in terms of salt removed by a unit volume of leaching solutions. Amount of Na removed from Shura soil by one pore volume of water, 0.05 N and 0.1 N H$_2$SO$_4$ (Figure 5.2) was 48 per cent , 60 per cent and 70 per cent of total sodium respectively. However, amount of sodium removed from Sabakh soil was 36.5 per cent, 40 per cent and 54 per cent of the total sodium respectively. Moreover, for both soils, increased amount of leaching solution percolating through resulted only in a slight increase in amount of Na leached. This may indicate that remaining amount of Na in soils takes fairly long time to be removed (Amezketa *et al.*, 2005; Lopez- Aguirre *et al.*, 2007).

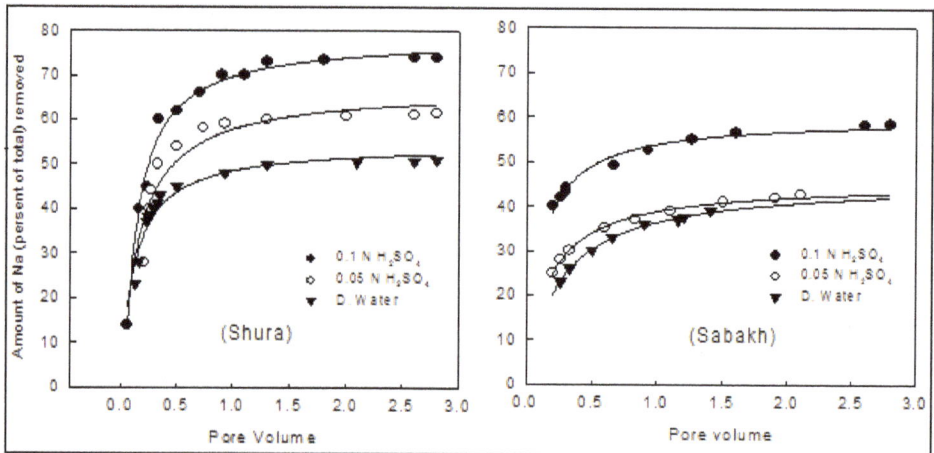

**Figure 5.2: Amount of Na (as a percentage of the total) Removed by different Leaching Solutions.**

Amount of $Ca^{2+}$ leached from shura and sabakh soils is shown in Figure 5.3. The greatest amount of Ca removed was under 0.1 N $H_2SO_4$ while the least amount was under the water. However, in sabkh soil only 85-90 per cent of Ca was leached under 0.1 N and 0.05 N $H_2SO_4$ while < 80 per cent was leached under water. Increasing the period of leaching to 3 pore volumes passed through resulted in only minute increase in the amount of $Na^+$, $Ca^{2+}$, or $Mg^{2+}$ leached.

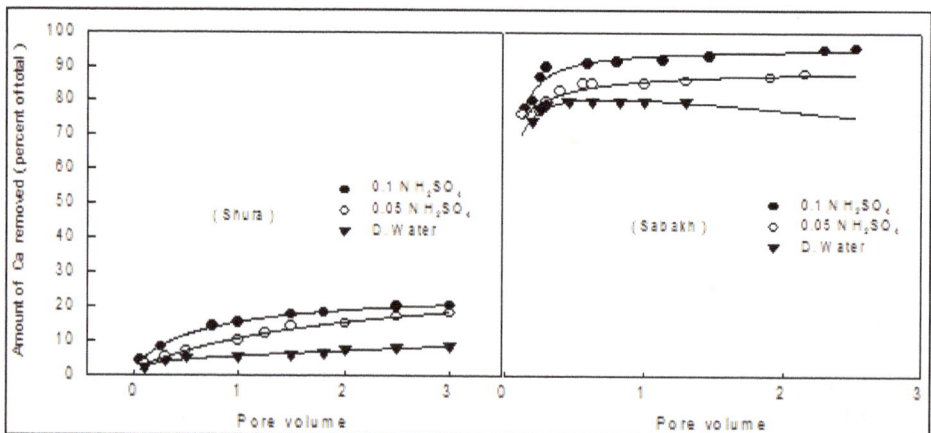

**Figure 5.3: Amount of Ca (as a percentage of the total) Removed by different Leaching Solutions.**

Amount of $Mg^{2+}$ leached from shura and sabakh soil (Figure 5.4) was relativity equal. Amount of $Mg^{2+}$ leached from shura soil was 98 per cent, 86 per cent, and 75 per cent under leaching with 0.1N $H_2SO_4$ 0.05N $H_2SO_4$ or distilled water respectively. This may again confirm the previous conclusion that solution of sulfuric acid is more effective in replacing $Na^+$, $Ca^{2+}$, and $Mg^{2+}$ ions than the water (Amezketa *et al.*, 2005; Lopez- Aguirre *et al.*, 2007). Amount of Na remained may behave as a seed for resalinization process.

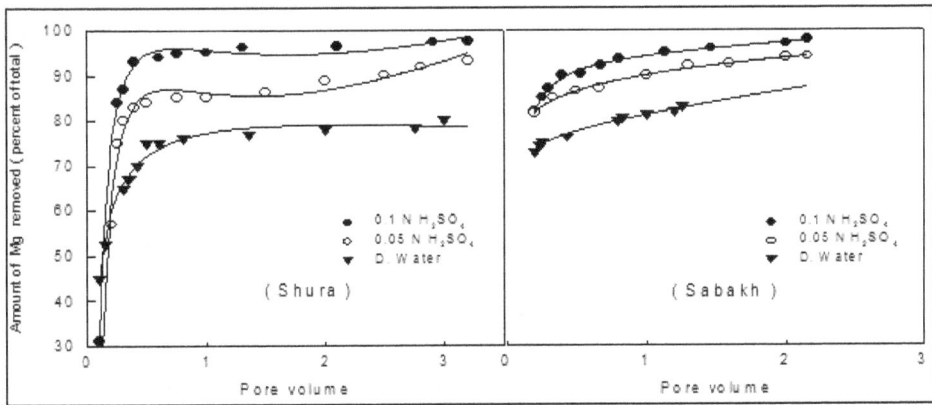

**Figure 5.4: Amount of Mg (as a percentage of the total) Removed by different Leaching Solutions.**

In the conventional reclamation process of salt affected soils the electrical conductivity of the leached and that of saturation paste are considered as the main parameter for efficient reclamation. In this study EC of the leachate was <4 dS/m yet there was >50 per cent of total $Na^+$ remained in both soils under the leaching with distilled water. However, amount of $Na^+$ remained in both soils was <30 per cent under the leaching with the acidified water (Ahmad *et al.*, 2013). This may again confirm the superiority of using $H_2SO_4$ as reclaiming aid over that of water. Moreover, leaching with acidified water may prevent reclaimed soils of being resalinized again.

## Summary

Leaching and hydraulic conductivity studies under laboratory conditions with dilute sulfuric acid solutions and water were carried out o two types of clayey salt affected soils namely Shura and Sabakh. The acid solution were quite effective than water for this purpose on both the soils, the efficiency being greater with increasing concentration of acid. Further, with increasing leaching volumes, the hydraulic conductivity was maintained in Shura soil, it declined greatly in Sabakh soil.

## References

1. Ahmad, S., A. Ghafoor, M. E. Akhtar and M. Z. Khan (2013). Ionic displacement and reclamation of saline- sodic soils using chemical amendments and crop rotation. *Land Degrad. Develop.* 24: 170- 178.

2. Al- Taie, A.H. (1970). Salt affected and water logged soils in Iraq. Report to seminar on method of amelioration of saline and water logged soils.

3. Amezketa, E., R. Aragues and R. Gazol (2005). Efficiency of sulfuric acid, mined gypsum, and two gypsum by- products in soil crusting prevention and sodic soil reclamation. *Agron. J.*, 97: 983-989.

4. Buringh, P. (1960). Soil and Soil Conditions in Iraq. Ministry of Agriculture, Republic of Iraq.

5. Kirda, C., D. R. Nielsen, and J. W. Biggar (1974). The combined effects of infiltration and redistribution on leaching. *Soil Sci. Soc. Ame. J.* 6: 323-330

6. Lopez- Aguirre, J. G., J. Farias- Larios, J. Molina- Ochoa, S. Aguilar- Espinosa, M. del R. Flores- Bello and M. Gonzales- Ramirez (2007). Salt leaching process in an alkaline soil treated with elemental Sulphur under dry tropic conditions. *World Journal of Agricultural Sciences* 3: 356- 362.

7. McNeal, B.L., D. A. Layfield, W.A. Norvell, and J. D. Rhoades (1968). Factors influencing hydraulic conductivity of soils in the presence of mixed salt solutions. *Soil Sci. Soc. Ame. Proc.* 32: 187-190.

8. Miyamoto, S., J. Ryan and J. L. Stroehlein (1975). Potentially beneficial uses of sulfuric acid in south- western agriculture. *J. Environ. Qual.* 4: 431- 437.

9. Page, A.L. (1982). 'Methods of soil analysis. Part 2, Chemical and Microbiological Properties'. 2nd edition (American Society of Agronomy Inc.: Madison, WI).

10. Papadopoulos, I. (1985). Soil salinity as affected by high-sulfate water. *Soil Sci. Soc. Ame. J.* 5: 376-381.

11. Quirk, J. P., and R. V. Schofield (1955). The effect of electrolyte concentration on soil permeability. *J. Soil Sci.* 6: 163-178.

12. Shainberg, I., M. E. Sumner, W. P. Miller, M. P. W. Farina, M. A. Pavan and M. V. Fey (1989). Use of gypsum on soils: areview. *Advances in Soil Science* 9: 2-111.

13. Shainberg, I., J. D. Rhoades, and R. J. Prather (1981). Effect of low electrolyte concentration on clay dispersion and hydraulic conductivity of a sodic soil. *Soil Sci. Soc. Ame. J.* 45: 273-277.

*Chapter 6*

# Agricultural Mechanization in Arid Regions of Developing Countries: Towards Enhanced Productivity

## C. Gomes, F.M. Hameed, N.N. Nasiruddin, M.M. Afiq and N. Abdullah

*Faculty of Engineering,*
*Universiti Putra Malaysia (UPM),43400 Serdang,*
*Selangor, Malaysia*
*E-mail: chandima@upm.edu.my*

## Abstract

Utilization of mechanized agricultural tools to upgrade farm operations in arid regions of developing countries is one of the best solutions available for global food crisis. Various efficient mechanized systems have been developed for agricultural applications in the last few decades. The mechanization is available for all phases of agricultural procedures; from soil preparation, up to gathering and storing of harvest. This study reviews the status of research and development in these techniques that are currently in use in the agriculture sector, with special attention to their applications in arid areas. Countries with arid or semi-arid regions could consider farm technique mechanization as a tool of augmenting existing methods to achieve high socio-economic goals. Mechanization optimizes labour management, cost and time of operation and quality of the yield. It also preserves two key resources of agriculture: cultivable soil and freshwater. Our study shows that there is a promising future for mechanization of agricultural techniques, especially in dry zones, to cater for the much-needed food supplies to the population in developing countries.

*Keywords*: *Agriculture, Mechanization, Arid region, Developing countries, Cultivable soil.*

## Introduction

There is a need for a change of existing agricultural techniques to resolve various issues that have emerged at the field-level in arid region of developing countries (Mada and Mahai, 2013). Agriculture is a progression of organized operations that is carried out by man on the soil and vegetation to produce edible products for raw materials and nourishment. Up to 50 per cent of the gross domestic products and materials to industries are provided by agriculture related processes (da Silva *et al.*, 2009). Thus, irrespective of the geographical location, agro-systems play a vital role in the global economy.

Territories in the world that have very little to almost zero rainfall are classified as arid regions. Without the help of rain, it is difficult to produce agricultural products in these areas. The places that are known for this blazing climate are mostly belong to the less-developed (also termed developing) world; most parts of Africa, Middle East, Northwest and South India, Northern China and Central Asia. Despite the arid nature of the regions, in many developing countries, agriculture helps in providing employment for most of the population.

Due to the unique concerns of agriculture in arid areas and the constraints and specific circumstances that prevail in developing countries, the agro-industry of less-privileged communities in dry landscapes needs special attention with respect to various aspects. To help in sustaining agriculture in these communities, farmers could be encouraged to adapt the practices of efficient techniques at various phases of agro-processes; seeding, fertilizing, harvesting, irrigation and drainage. Mechanization is one such vital focal point in this regard. It is the process of shifting from working largely or exclusively by hand or with animals to doing the same with machinery. During the last few years there are some significant improvements in the technology with respect to the above subject. There are few reports on these latest developments (Onwude *et al.*, 2016), however, there is a lack of literature that focuses on agricultural mechanization in arid regions. Such review has become a substantial need at present. This work is an attempt to fill-up that scientific gap.

## Information Analysis

### a. Characteristics of Arid Areas

There are several types of climate classification as per various aspects of the climate parameters; Aridity index, Alisov climate classification, Köppen climate classification, Holdridge life zone classification, Trewartha climate classification, Vahl climate classification (Arora, 2002). Among them Aridity Index (AI) plays a key role in identifying climate based regional characteristics which are important for agriculture. An aridity index is a numerical indicator of the degree of dryness of the climate in a given region. Several aridity indices are in use. Each AI characterizes, locates or delimits regions that experience deficit of available water. Thus, AI is directly applicable in identifying the impact of water availability for agriculture or livestock industry. As per the more recent definitions, AI is computed as the ratio between the average annual precipitation and the potential evapotranspiration of a given location (UNEP, 1992). The Table 6.1 depicts arid regions as per this AI definition.

**Table 6.1: Classification of Arid Regions as per AI Definition by UNEP (UNEP, 1992)**

| Classification | Aridity Index (AI) | Global Land Area |
|---|---|---|
| Hyper-arid | AI < 0.05 | 7.5 per cent |
| Arid | 0.05 < AI < 0.20 | 12.1 per cent |
| Semi-arid | 0.20 < AI < 0.50 | 17.7 per cent |
| Dry sub-humid | 0.50 < AI < 0.65 | 9.9 per cent |

Table 6.1 shows that 47.2 per cent of the world land area belongs to some type of arid climate. Almost $1/5^{th}$ of the world land area is arid or hyper arid and another similar percentage is semi-arid. This data clearly depicts that technology development for dry climate agro-industry is a dire need for the sustenance of human food supply. Figure 6.1 shows a semi-arid landscape in South Africa, which could be considered as a cultivable land under appropriate agro-technology.

**Figure 6.1: Semi-arid Landscape in South Africa.**

## b. Mechanization

Mechanical inputs in agricultural sector could broadly be categorized into 3 types; hand-tool technology, draught-animals, power and engine-power technology (Odigboh, 1999). These categories are also known as levels of mechanization.

The first level of mechanization, hand-tool technology, relies on human labour as the source of power. This refers to the usage of tools and implements like blades, peaks, machetes, hoes, spades, forks, rakes, axes, knives, manually powered winnowers and seed drills, which uses the human muscles as the source of its power (Sims and and Kienzle, 2016). However, there is a limited power output in human labour although the demand for power work in agricultural operations is high. The time taken to carry out a certain quantity of work in different farming operations varies according to various considerations such as the type of crop, soil moisture, optimum seeding dates as well as the soil type. Not only their health plays a major impact on the amount of time and quality of their work, but also the climatic conditions especially in their regions where it is high ambient temperature and humid. These factors could affect the capacity of their work, as the use of manual labour is a menial work and it limits timeliness of operations (Bellon, 2001; Aikins, 2000).

The second level of mechanization is the usage of draught animal in agriculture. Animals such as horse, oxen, mule, donkey and camel are often used to power the machines and equipment in agriculture (Sims and and Kienzle, 2016). The optimum pulling force that could be powered by horses, donkeys, mules and camels are about 12 per cent to 15 per cent of their body weight whereas the working speed for most of these draught animals at their optimum pull is 3.5 km hr$^{-1}$ (Ashburner *et al.*, 2009). Similar to human labour, the general conditions of these animals and the climatic conditions are the main factors in defining the portion of work they are capable to carry out. The poor conditions of draught animals that can be observed especially at the end of a dry season, is another point of concern. This becomes a typical issue, in the cases where farmers rely purely on grazing for these animals to feed on. The grasses and other fodder plants that finally dry out are the least nutritional and plentiful. Thus, draught animals would lose weight and strength that could lead to declination of working capacity.

The final level of mechanization is the most commonly employed source in the field of agriculture; usage of engine power technology. Engine-powered technology takes various forms that include a wide range of tractors and accessories that are mainly used as a mobile power on the field of operations, engines and motors that use petrol, diesel or electricity to power up the mills, irrigation pumps or threshers, and self-propelled machines in seeding, harvesting and production (Sims and and Kienzle, 2016). These mechanical powers enable the farmer to manufacture a larger production compared to using manual labour and draught animals. Basic engine-powered agro systems include basically; tractors (track type and field type), walking tractors and combines or mower. There are many accessories available to be fitted into these machines which make them applicable for multi-purpose tasks. Figure 6.2a depicts a two-wheeled walking tractor which can be used for multipurpose tasks; tilling, plowing, winnowing and even generating electricity in need and Figure 6.2b a walking tractor with a trailer and driver seating facility which is mainly used for transportation of goods and even passengers. These machines are extremely useful in farmlands in arid and semi-arid regions in developing countries due to their very efficient fuel consumption capacity.

**Figure 6.2a: A Walking Tractor.**

**Figure 6.2b: A Walking Tractor with a Trailer and Driver Seating Capacity.**

It should be emphasized that modern mechanization does not mean that the process essentially needs external energy (fossil fuel, electricity *etc.*). For example, a hand operated winnowing fan, improvised with a gear wheel system for higher efficiency is a "good" mechanization.

## c. Mechanization Potential in Land Preparation Process

Land preparation is a basic activity in most agricultural processes. It is the initial process immediately before the crop planting stage. Process of land preparation usually starts with clearing the potential agricultural land from previous crop residues or unwanted vegetation, and then followed by soil tillage (FAO, 2015). The land preparation is essential to foster the required soil conditions that prevent soil erosion, soil surface crusting and soil water wastage (Klein and Zaid, 2002), apart from uniform soil fertility distribution and suppressing of weed germination. There are several factors that need attention in land preparation process such as irrigation water quality and supply, land selection, mechanical operations to be implemented, tools and equipment, labour and time schedule (Ofori, 1993).

Arid regions experience long term dryness due to the climate and insufficient water supply. FAO (1989) has classified three arid zones (Table 6.1), which are hyper-arid zone (arid index 0.03), arid zone (arid index 0.03-0.20) and semi-arid zone (arid index 0.20-0.5). Aridity index can be calculated by precipitation divided by potential evapotranspiration. The differences between normal condition and arid condition, required for the planning of an agricultural project, are temperature, atmospheric and soil humidity, wind, rainfall rate, and various soil properties such as salinity, acidity, porosity, compactness *etc.* These factors are critical challenges for agricultural engineers and farmers, in the process of overcoming the issues in planting and maintaining crops in arid regions.

Tillage is the main process to be focussed on in land preparation. Tillage can be defined as manipulating the physical structure of soil by loosening, crushing or compacting to optimize the crop germination or seedling growth. The objectives of tilling the soil are seedbed preparation to facilitate seed planting, reducing undesired levels of soil erosion and weed control. Such objectives, effectively achieved, will help farmers to increase the rate of successful seedling growth (Dumanski, and Peiretti, 2013). Tillage also affects soil aeration that influences soil moisture and water infiltration; however, the results depend on the soil type and its composition and the content of organic matter in soil (Rusu *et al.*, 2009). Nicou and Charreau (1985) reported increasing yield due to tillage effect on different crops as it is given in Table 6.2. The outcomes show that in arid regions, tillage plays a crucial role in increasing crop yields. These results have repeatedly been verified over the years by many researches (Afzalinia *et al.*, 2011; Alavijeh *et al.*, 2014).

**Table 6.2: Data of Increasing Yield by Tillage in West African Arid Tropics (Nicou and Charreau, 1985)**

| Crop | Yield kg/ha | | Increase of Yield |
|:---:|:---:|:---:|:---:|
| | *Without Tillage* | *With Tillage* | |
| Rice | 1164 | 2367 | 103 per cent |
| Corn | 1893 | 2791 | 50 per cent |
| Sorghum | 1691 | 2118 | 25 per cent |
| Groundnut | 1259 | 25 | 24 per cent |
| Millet | 1558 | 1894 | 22 per cent |
| Cotton | 1322 | 1550 | 17 per cent |

Historically, tillage has hardly been applied in land preparation in arid areas of Africa. For an example, even today, farmers in Zimbabwe, use traditional tools such as hoe, sharp pointed stick, mouldboard plow and manual harrow for preparing their farm. These tools were manually hand held and some were used with animal power. This practice yields poor soil fertility, high losses of soil, nutrients losses and low soil water holding and finally, low yield (Biamah, 2007; Elwell, 1993). In semi-arid regions of many East African countries iron plow is used to till the soil without any machine power or even animal power. In West Africa too, farmers practice hand cultivation in traditional systems of cultivation. Animal powered traction has been introduced in early 20th century and was developed gradually by time and area. In India, ox traction is commonly used for plowing and land preparation in some arid regions for over few millennia.

Many agronomists state that deep-tillage of plowing area is the best for soil and crops, however, these methods could lead to soil erosion and rapid organic matter decomposition in the long run (Haddaway *et al.*, 2016; Jin *et al.*, 2011). Thus, optimum tillage should always be implemented considering several technical factors such as soil physical properties, outcomes with and without tillage for few seasons and soil management techniques. Furthermore, suitable machinery needs to be used on tillage such as tractor to increase tillage efficiency and reduce manual labour. Many tools, machines or tractor types can be used for tillage operation with various components that serve different purposes; chisels, light and heavy discs, single, twin and multiple blades or discs, mouldboard plows, harrows and rotary tillers.

Tillage can be classified as conventional tillage, conservation tillage and no-tillage. Conventional tillage is a method that basically applied for loosening the soil by using certain machines or tractor implements like plow, disc and cultivator. Conservation tillage is a technique for planting seed that minimizes the disruption of soil by using specific machines or tractor implements as planned. Conservation tillage systems have been proven to be effective in increasing crop yield besides maintaining deformable soils. In semi-arid region, tillage operations, such as seedbed preparation, are done by tractors and mechanized equipment such as disc plows and chisels implements. Cultivation processes in arid and semi-arid region are very much dependent on soil properties and moisture content. This operation needs an appropriate operation and time plan for tillage as the limited water supply may make the agricultural land too dry to cultivate. Thus, the concept of conservation tillage should essentially be put into practice in drylands taking the pertinent soil properties and processes into account (Afzalinia *et al.*, 2011).

The most detailed research up to date on the types of tilling method on arid landscapes has been reported in Alavijeh *et al.* (2014). Their results, on two maize varieties, depict that tillage methods have significant positive effects on seed weight and number of grains per year, wet weight of leaf, stem and grain quality traits such as soluble sugars and protein, soil physical properties. They have tested four types of tilling techniques: tillage with rotary tiller, cultivator with blade and light disk, twin disk, and mouldboard plow and light disk. Interestingly, the outcome shows that each tilling method has its own merits and demerits.

There are many support practices in land preparation operations such as contouring, furrow diking, strip cropping and terracing, which need farmer's attention during the agricultural operations. These techniques are usually used in parallel with tillage for aiding soil and water conservation ability (Khlifi *et al.*, 2010; Unger *et al.*, 1991). Figure 6.3a shows a tractor fitted plough which is designed to

**Figure 6.3a: A Plough Attached to a Tractor.**

**Figure 6.3b. A Mechanized Drag (Photos adopted from "Didactic Material: Modern machines in agriculture").**

open furrows in the earth and Figure 6.3b a mechanized drag which could break up the earth, especially in drylands with hard soil (soil with high compactness).

## d. Irrigation and Drainage

Agricultural mechanization also includes irrigation and drainage systems together with related technologies and equipment. Applying the required amount of water uniformly at the right time is important for getting a high yield in agricultural fields. If water could be applied precisely in the right amount, it will improve the quality and quantity of the production while optimizing the available water resources. The traditional exercise of agricultural irrigation in Asia and South America is flood irrigation, which demands ample source of water. Most parts of Africa were also used to practise this method but acute shortage of water in the region prompted them searching for alternative techniques. As a result, in the recent past, centre pivot irrigation systems (Almasraf *et al.*, 2010; Waller and Yitayew, 2016) have become increasingly popular in semi-arid regions of the African continent (Figure 6.4). In several countries, such as Zambia, centre pivot irrigation systems have now outnumbered subsurface drip irrigation and other types of sprinklers (Mendes *et al.*, 2014).

**Figure 6.4: Centre Pivot Irrigation System.**

A centre pivot irrigation system consists of a pipe and framework (span) supported by two wheeled-towers. The drive mechanism of the tower could be automated, remote controlled or manually operated. Pipes have outlets which are fitted with sprinklers, known as emitters. Emitters could be elongated (if necessary) by fitting them to the outlets via rigid or flexible droppers. Many new modules have length adjustable droppers so that the water sprouting height could be varied from plant to plant (or the same plant at different growth heights). Systems that have been used in large rectangular fields have a lateral movement system as well to cover a sizable area effectively.

Proponents of centre pivot irrigation systems claim that the technique has the following edge over the other irrigation mechanisms.

## The System

1. Could easily be fully automated based on atmospheric and soil conditions.
2. Could be programmed for reliable performance with state-of-the-art controls for precise irrigation management and scheduling.
3. Could be remote controlled and monitored via Bluetooth (short range) or internet and mobile phone (long range).
4. Has a versatile range of applications; multiple crops under one circle and the same emitter spacing for different crops.
5. Has proven record of above 90 per cent efficiency and uniform water distribution under a range of climate, wind, soil and slope (up to 30 per cent inclination) conditions.
6. Reduces deep percolation, salt build-up and runoff of chemicals into the neighbourhood.
7. Has low capital investment per unit area.
8. Is eco-friendly due to the recyclable components, over 20 years of continuous operation, easy removability from the field and nontoxic materials.

In arid regions that have very little to no rainfalls to water the ground, micro-level irrigation seems to be the best plausible solution that can be used to increase the crop yield with limited water resources. It opens many possibilities to invest in mechanization. In many developing countries, especially those in Africa, one could clearly notice the deficiencies in government planning that facilitate or encourage mechanization in micro-scale irrigation and drainage. Many countries have over-reliance on unsuitable and unpredictable external mechanization inputs. For an example, in many arid parts of Africa, large-scale investments by respective governments are very scarce on field-level agriculture-related irrigation which plays an essential role in magnifying crop production. This is mainly due to the fact that investments in mechanization are being vested only on large commercial farms or government schemes. Even when funding is allocated for micro-scale irrigation schemes, it is frequently observed that they are short-term investments where managements are incompetently trained and stakeholders are not well-informed regarding the project objectives and roadmap. The success of investment on the mechanization of micro-scale agro-sector depends also on how well the stakeholders understand the backdrop of the socio-economic system; the structures of landholding and landownership, fiscal regime, financial control, operational responsibility distribution, education and training, research and development. It should be noted that only 17 per cent of the croplands with irrigated agriculture, which bring about only 40 per cent of the world's food (FAO 2014). This small percentage of irrigated land could be attributed to the lack of investments in irrigation in the agriculture in arid/semi-arid regions in the developing world, especially in Africa, compared to the same in developed countries such as the USA, Canada and Europe. As an example, in Africa, only about 5 per cent of the arable land is irrigated whereas the figure in India, another developing country in Asia, is almost 40 per cent (FAO 2014).

Irrigational requirements of a given agro-field are strongly dependent on the pre-conditioning of land and methods of farming. Thus, methodical irrigation generates parallel mechanization opportunities, especially in smallholder farms; *e.g.* adaptable engines fitted in tractors and power tillers. If the operation of the machinery and equipment is carried out effectively and efficiently and is combined with high-yielding varieties, essential fertilizers and chemicals in minimum required amounts and appropriate disease control, irrigation could have a major impact on water usage, which is very beneficial to harvest many farm productions with high yield (Houmy *et al.*, 2013).

Nonetheless, machinery and equipment used in irrigation systems can present both direct and indirect ecological dangers in several ways. For an example, if the earthmoving and other machinery/equipment in the construction of irrigation systems are used negligently, it may seriously defile the ecosystem. Other than that, inappropriate usage and management of the operation of pumps and other on-farm equipment may stimulate levels of agricultural augmentation that leads to soil degradation and salinity.

Having a proper irrigation system alone is not sufficient for successful farming, if the land has a poor water management system and deficient drainage infrastructure. The main problem that may arise due to bad water management is waterlogging in the irrigated areas. Consistent build-up of salt concentrations is the cause to waterlogging and it will eventually lead to soil degradation. By the end of the last millennium, the percentage of agricultural lands in under-developed countries was 7 per cent (Smedema *et al.*, 1998). The availability of arable lands in these countries is declining over the years (Hallam, 2009). One can observe some form of agriculture related drainage systems only in few of these countries. Half of the world's irrigated land experiences drainage problems, which may be the reason of the 25 million hectares of land in the world being sterile. Added to this situation, a land mass of 0.5 million hectares per year is becoming unsuitable for agriculture due to waterlogging and salinity (Smedema, 2000). According to the reports of Department of Agriculture (USA), in Mid-West of United States, the modern drainage in the arid and semi-arid regions of the country started in the 18th century and then expanded to the west parts of the country that is termed arid by the advent of the 20th century. In maintaining and boosting the productivity of their croplands, these developed countries with arid/semi-arid landscapes invested in large-scale irrigation projects well in advance realizing the importance for drainage in an expanding agro-sector. Unlike agro-irrigation, which could successfully be applied at micro-scale, agro-drainage should be addressed at macro and mega scales for better results. It could be proposed that international funding bodies should escalate their support in developing proper drainage systems in developing countries, giving special attention to hyper-arid regions in countries such as Libya, Sudan and Pakistan.

Drainage benefits many sectors; economy, environment and public health, if it is done properly. In the sector of economy, the rural welfare gains the advantage from having proper drainage as it provides the continuous natural resources that lead to the enhancement of national economy. In terms of the environmental benefits, it eases the farmers from having infertile lands as drainage extracts the surplus moisture and salts directly in the root zone. Consequently, this leads to a productive environment for any other agricultural plantation. Improvement of drainage systems could increase production yield between 10 per cent and 20 per cent in the case of 13 different crops (Qadir *et al.*, 2014; World Bank, 1991). In terms of benefits in public health, drainage eliminates the risk of waterborne and vector-borne diseases by controlling waterlogging and salinity.

There are various practices in drainage:

1. Open drains; the most famous practices around the world,
2. Subsurface drainage; practiced either through implementing vertical drains (tube-wells) or horizontal subsurface drains (buried pipes),
3. Pumped tube-wells; most commonly used in places with substantial amount of fresh groundwater aquifers that is usually for irrigation and
4. Pipe drainage; implemented in most lands in arid regions in developing countries.

The appropriate type of drain has to be determined by cost-effectiveness, budget of the construction and cost of operations and maintenance. Other than that, the crop type and its ferocity, the features of the water of the ground, the technical extent to carry out the structural work and the market prices and economic return should also be identified before selecting the right type of drain.

In summary, a case-designed and well-planned mechanization of irrigation and drainage is crucial in having a steady growth of irrigated agriculture in arid regions. Such system, once well-maintained and well-managed will prevent the ground from waterlogging and buildup of salinity. With the right and appropriate mechanization, farmers in the arid regions of developing countries will be able to feed not only their family but also the entire population. By augmenting and optimizing their farming techniques and methods through various levels of mechanization, the entire agro-sector in these less-privileged communities would ultimately be led to prosperity with improved land use, increase in food production and on a nation scale, a bigger export potential.

### e. Seed Planting and Fertilizing

Seed planting is one of the most labour-intensive work phase in the traditional agricultural practices. In arid regions, due to the harsh weather conditions and hard surface soil, manual seed planting (without tool support) is an exhaustive and time-consuming process which result low rate of seed placement, poor spacing efficiency and serious back pains and other muscular ailment (both in the short and long runs) at the worker's end. Thus, optimization of seed planting through mechanized technologies is a dire need in such parts of the world.

There are many mechanized seed planting devices, popularly known as seed planting machines have been developed during the last decade (Adisa and Braide, 2012; Kyada and Patel, 2014). Among them, human operated machines are basically recommended for small scale farms in arid regions. These machines usually require well-prepared seedbed either ridged or flatbed type. Basically, the seed planting machine should place the seed and fertilizer, typically in rows at pre-determined depth and seed spacing. Then the seed should be covered by compact soil. The applied compactness of covering soil is a sensitive parameter to be pre-determined in the machine design stage. Most of the above parameters are seed type, soil type and climate dependent variables which makes a machine developed for planting a certain variety of seed in a given environment not applicable, as it is, in another case.

Kyada and Patel (2014), points out the following mechanical factors of a seed planting machine should be optimized to achieve the best performance.

1. Uniformity of depth of placement of seed.
2. Even distribution of seed along rows.
3. Transverse displacement of seed from the row.
4. Prevention of loose soil getting under the seed.
5. Uniformity of soil cover over the seed.
6. Mixing of fertilizer with seed during placement in the furrow.
7. Seed metering system that meter the seed from the seed box and deposit it into the delivery system

A hand-operated seed planting machine developed by Kyada and Patel (2014) has been depicted in Figure 6.5. The device is well suited for agro-fields in semi-arid regions.

Tractor-integrated machine planters are more suited for large scale agricultural fields to save labour and time. Most often these machines are multifunctional (known as combined machines or combines), so that one tractor with changeable accessories could act as a combined farming and fertilizer spreading machine (Rathke *et al.*, 2006; Chaifai Elalaoui, 2008). Typically, such system has a small four-wheel tractor, a bracket, a seed and fertilizer bucket, a sowing and fertilization device and suction pump (optional). The tractor is fitted with grain sowing machine and rotary machine. The motor is driven by the battery of the tractor. Sowing and fertilization central discs can be selected according to the grain or seed size, seeding density and fertilization amount. These multifunctional combine farming and fertilizer spreading machines are extensively used in corn cultivation in several arid regions of Africa at present (Orth, 2009). Figure 6.6 shows a tillage planter: a machine that is able to plant seeds without prior tilling. These types of combines are extremely helpful in cultivating drylands with hard soil where labour for tilling is either insufficient or too expensive to afford.

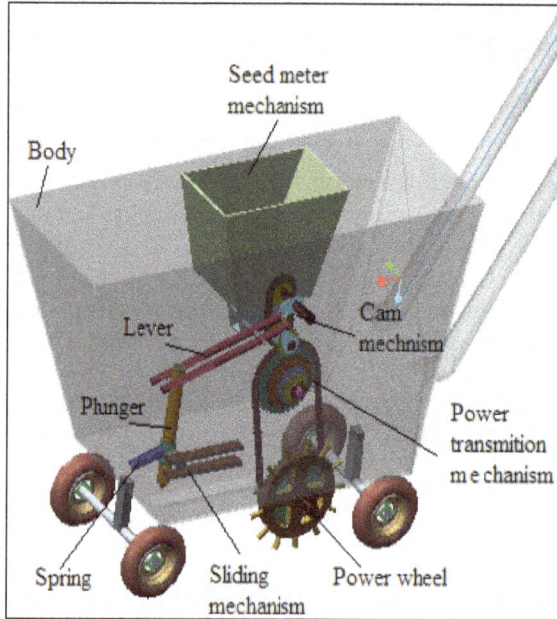

**Figure 6.5: A Hand-Operated Seed Planting Machine Developed by Kyada and Patel (2014).**

**Figure 6.6: A Tillage Planter.**

## f. Harvesting Technologies

Harvesting is another labour intensive operation in traditional agricultural cycle, which is required to be done at proper time to secure a good yield. Postponement in yield gathering will have a direct impact on quantity and quality of products. Furthermore, the method of harvesting effects on wastage proportion. Traditionally, harvesting in developing countries depends on human labor and particularly in small land holdings. Nonetheless, in such situation where labor shortage is on the rise, mechanical harvesting has become inescapable. Mostly, manual harvesting of all crops is done by sickle which is arduous and time consuming. Harvesting and threshing are the most significant processes in the field operations, which are laborious involving human effort (Veerangouda *et al.*, 2010).

Harvesting of wheat crop, for an example, in a short possible time after maturity is necessary to decrease shattering losses and delay plant next crop. Further, the natural extreme events such as heavy rain, hailstorms and windstorms that occur in the harvesting season may incur heavy losses. Utilization of reaper plus thresher or combine can resolve the labour shortage as these harvesters can sever and thresh the crop simultaneously. Chaudhry (1979) evaluated 2.0 per cent and 1.2 per cent grain losses on account of tractor threshing and combine harvester, respectively. Mechanical harvesting not only lessens the post-harvest losses but also aids in shortening the harvesting interval. Grain losses of wheat when harvest and thresh manually also have been determined to be 4.1 per cent by Bala *et al.* (2010). Shamabadi (2012) estimated the performance of eight combines and found that time of harvesting, field topography, relative humidity, seed moisture content, and varietal characteristics are the essential factors affecting harvest losses. He determined that mean total grain losses by different combines was 6.88 per cent at wheat harvesting stage. Sattar *et al.* (2015) measured grain losses of wheat by different harvesting and threshing techniques. They observed that losses in all fields of study were 4.28 per cent, 3.85 per cent, and 2.92 per cent of the total yield during harvesting and threshing operations with manual and thresher; reaper and thresher; and combine harvester, respectively. According to Begum *et al.* (2012), grain losses of wheat during the threshing activity are observed mainly in the form of broken grains. The economic advantages of manual harvesting plus mechanical threshing and combine harvester have been studied by Razzaq *et al.* (1992). They concluded that combine harvester has higher wheat yields than manual harvesting plus mechanical threshing.

Combine harvester demonstrated more economical than manual harvesting plus mechanical threshing actually in developing countries. Pawar *et al.* (2008) observed the total losses of combine harvester are 4.2 per cent, whereas combination of reaper with thresher resulted in 10.6 per cent loss. In this study, the harvesting costs by these two techniques also have been calculated, where they manifested the operational cost of combine harvester is 50 per cent less than the other combination. Besides that, they proved both techniques can be employed appropriately for large and small fields. In a study which set out to determine Hossain *et al.* (2015) found that the overall field utilization efficiency of combine harvesters is 97.5 per cent, whereas the same in manual harvesting is 35 per cent. They considered power exploitation, cost and time in calculating the efficiency.

As per the above outcomes, one may suppose that combine harvester is an efficient, economical, and less labour demanding machine. It increased grain retrieval by reduction harvesting and threshing losses.

## g. Crop Management and other Concerns

Crop management is a term with wide range of implications; keeping up with latest technologies in crop selection, seeding, land preparation, watering, fertilizing, weed and pest control, plant health care and yielding. In that sense, it covers almost the entire spectrum of cultivation. Good crop management policy considers the consequences of faming methodologies on the sustainability of agro-sector in the long run. Process mechanization could be introduced into this management concept at various stages of the agro procedures, as it has already been discussed. However, no mechanization effort will be fruitful unless the entire crop management scheme is in the right track. Hence it is appropriate to wind up this work by addressing this important overall management procedure in agriculture.

In the recent years, integrated crop management (ICM) concept turned out to be the most productive method in many developed countries. The ICM scheme has been built on the objective of reducing external farm inputs, such as inorganic fertilizers, pesticides and fuel by better management of inputs and/or replacing them with farm produced substitutes. These changes need to be introduced without significant negative impacts on the crop yield. In fact, ICM could considerably reduce the farming cost, environmental pollution and agricultural sustainability in the long term. A close analysis shows that the concept is ideal for agriculture management in arid/semi-arid regions. ICM could be optimized by adopting site-specific procedures in the following phases of cultivation.

1. Decision taking on crop rotations
2. Selection of the most-fitting cultivation techniques
3. Selection of the appropriate seed varieties and method of seed planting
4. Level of reliance on artificial inputs such as fertilisers, pesticides and fossil fuels
5. Maintenance strategies of the landscape
6. Post-harvest activities (storage, transportation, recycling, revamping)
7. Conservation of neighbouring wildlife habitats

The above steps of ICM facilitate the farmland in enduring long-term sustainability, especially under harsh environmental conditions. As agricultural mechanization is introduced under ICM, one should carefully pay attention to the negative impacts of such adaptation as well. Although, man-powered and animal-powered mechanical improvisations are possible, modern mechanization most often accompany external energy requirements. Such mechanisms demand proper system management (operational safety, storing/parking facilities, equipment health, repairing and replacements) and technical expertise. Furthermore, in developing countries, engine-powered mechanization is not locally made. Hence, it could be expensive for the farmer and loss of foreign exchange for the government. The maintenance of the engines and the repairs could also be a drawback, as operational skills are habitually underdeveloped and scarce in rural areas. The fuels and lubricants that need to run the engines are also high-priced in many developing countries.

Engine-powered mechanization replaces human labour that depends on agriculture as their source of income, thus shifting to such scenario may cause social issues as well. One may also argue that due to the requirement of operational overseeing, maintenance, repairing *etc.*, a new line of employment could also be generated with engine-powered mechanization.

One way out for the fossil fuel requirement is to use solar power for engine operations. In most arid regions, solar energy is amply available throughout the year, thus solar photovoltaic energy could easily replace fossil fuels in almost all mechanical systems. Solar power could also be used for running post-harvest storage facilities, which is a dire need in many arid regions of the developing countries. For an example, in Africa, one of the staple food in most parts is corn, which generates income to millions in agricultural industry. The poor post-harvest storage facilities lead to the loss of 14 per cent - 36 per cent of corn grains that the continent produces (Tefera *et al.,* 2016). Such wastage is intolerable to a region where hunger is a key enemy among the entire population.

## Conclusions

These findings have significant implications for the understanding of how agricultural mechanization can help getting better rural livelihoods in arid regions of developing countries. Mechanization technologies play a major role in increasing productivity and rural income while reducing the drudgery correlated with manual tasks. Machine applications in agricultural activities give benefits through all stages of farming processes; from soil preparation until post-harvest activities. As discussed in this work, we firmly recommend that mechanization as a prime contender for achieving agriculture related socio-economic goals in arid regions of developing countries. We also suggest implementing mechanization within an integrated crop management framework. In such case, one can pay attention on reducing the drawbacks of mechanization to achieve sustainable agro-economic goals in the long run.

## Acknowledgements

The authors would like to thank all departments of the Faculty of Engineering, Universiti Putra Malaysia in providing required facilities to make this work a success.

## References

1. Adisa A. F. and F. G. Braide (2012). Design and development of template row planter. *Transnational Journal of Science and Technology*, 2, 7, 27-33, August.

2.  Aikins, S. (2000). Animal drawn implement development for land preparation and moisture conservation. *M.Phil. Thesis*, Cranfield University, UK.

3.  Alavijeh, H. R. G., H. A. Chenarbon, B. Zand and M. Hamidi (2014). Effects of different tillage methods on soil physical properties, grain and forage yield of two cultivars maize. *Academia Journal of Agricultural Research* 2(1): 008-015.

4.  Afzalinia, S., A. Karami, M.H. Talati, and S.M. Alavimanesh (2011). Effect of tillage on the soil properties and corn yield. CSAE Paper No. 11-204, 10-13, Winnipeg, Manitoba, July.

5.  Almasraf S., J. Jury and S. Miller. (2010). Field evaluation of center pivot sprinkler irrigation systems in Michigan. Internal Report, Department of Biosystems and Agricultural Engineering, Michigan State University, USA.

6.  Arora V. K. (2002). The use of the aridity index to assess climate change effect on annual runoff. *Journal of Hydrology*, 265: 164–177.

7.  Ashburner, J. and J. Kienzle (2009). A review of some public sector driven mechanization schemes and cases of private sector models in Africa. In: Investment in agricultural mechanization in Africa. *Agricultural and Food Engineering Technical Report* No.8, FAO, Rome, 2011. pp. 51-58.

8.  Bala B. K., M. A. Hoque, M. A. Hossain and S. Majumdar (2010). Post-harvest loss and technical efficiency of rice, wheat and maize production system: assessment and measures for strengthening food security. Final Report (CF # 6/08), National Food Policy Capacity Strengthening Programme (NFPCSP), Ministry of Food and Disaster Management, Dhaka, Bangladesh.

9.  Begum, E.A., M.I. Hossain and E. Papanagiotou (2012). Economic analysis of post-harvest losses in food grains for strengthening food security in northern regions of Bangladesh. *IJAR-BAE.* 1(3): 56-65.

10. Bellon M.R. (2001). Participatory research methods for technology evaluation: A manual for scientists working with farmers. International Maize and Wheat Improvement Center (CIMMYT), Mexico.

11. Biamah E.K., G. Sterk, and L. Stroosnijder (2007). Tillage and Farmyard Manure Effects on Crusting and Compacting Soils at Katumani, Semi-arid Kenya, Discovery and Innovation, 19 (4), 254-263.

12. Chafai E. A. (2008). Sustainable agricultural management of drylands, International Centre for Advanced Mediterranean Agronomic Studies (CIHEAM), Near East and North Africa.

13. Chaudhry, M.A. (1979). Wheat losses at the threshing and winnowing stages. *J. Agri. Mech. in Asia, Africa and Latin America.* 10(4): 67-70.

14. da Silva C. A., D. Baker, A. W. Shepherd, C. Jenane and S. Miranda-da-Cruz (2009). Agro-industries for Development, The Food and Agriculture Organization of the United Nations.

15. Dumanski, J. and R. Peiretti (2013). International Soil and Water Conservation Research, Modern concepts of soil conservation, 1, 1, 19-23.

16. Elwell H. A. (1993). Development and Adoption of Conservation Tillage Practices in Zimbabwe. Soil Tillage in Africa: Needs and Challenges. Natural Resource Management and Environment Department FAO.

17. FAO (2014). World Agriculture: Towards 2015/2030, An FAO perspective, Corporate Document Repository, Economic and Social Development Department.

18. FAO (2015). Machinery, Tools and Equipment, 2014. Corporate Document Repository, Economic and Social Development Department.

19. FAO (1989). Chapter 1: The Arid Environment. Arid Zone Forestry: A Guide for Field Technicians, FAO Corporate Document Repository, Economic and Social Development Department.

20. Haddaway, N. R., K.Hedlund, L.E. Jackson, T. Kätterer, E. Lugato, I. K. Thomsen, H. B. Jørgensen, and P. E. Isberg (2016). How does tillage intensity affect soil organic carbon? A systematic review protocol, Environmental Evidence, 5: 1.

21. Hallam D. (2009). Foreign investment in developing country agriculture – Issues, policy implications and international response. VII Global Forum on International Investment, Session 2.2: Promoting responsible international investment in agriculture, December.

22. Hossain M. A., M. A. Hoque, M. A. Wohab, M. A. M. Miah, M. S. Hassan (2015). Technical and Economic Performance of Combined Harvester in Farmers' Field. *Bangladesh J. Agril. Res.* 40(2): 291-304.

23. Houmy, K., L. J. Clarke, J. E. Ashburner, and J. Kienzle (2013). Agricultural Mechanization in Sub-Saharan Africa: Guidelines for Preparing a Strategy. Rome: Food and Agriculture Organization of United Nations.

24. Jin H., L. Hongwen, G.R. Rabi, W. Qingjie, C. Guohua, S. Yanbo, Q. Xiaodong, and L. Lijin (2011). Soil properties and crop yields after 11 years of no tillage farming in wheat–maize cropping system in North China Plain. *Soil Till. Res.* 113: 48–54.

25. Klein P. and A. Zaid Date (2002). Chapter 6: Land Preparation, Planting Operation and Fertilisation Requirements. Agriculture and Consumer Protection, FAO Corporate Document Repository, Economic and Social Development Department.

26. Khlifi S., H. Arfa, L. Ben D. D'beya, S. Ghedhoui and S. Baccouche (2010). Effects of Contour Ridge Benches on Several Physical and Chemical Soil Characteristics at the El Ghrifettes Site (Zaghouan, Tunisia), Arid Land Research and Management, 24, 3.

27. Kyada, A. R, and D. B. Patel (2014). Design and development of manually operated seed planter machine, 5th International and 26th All India Manufacturing Technology, Design and Research Conference (AIMTDR 2014), IIT Guwahati, Assam, India, December.

28. Mada D.A. and S. Mahai (2013). The role of agricultural mechanization in the economic development for small scale farms in Adamawa State, The *International Journal of Engineering And Science (IJES)*, 2, 11, 91-96.

29. Mendes D.M., L. Paglietti, D. Jackson and F. Chizhuka (2014). Zambia: Irrigation market brief, FAO/IFC Cooperation Programme, Economic and Social Development Department.

30. Nicou R. and C. Charreau (1985). Soil tillage and water conservation in semi-arid West Africa. In: *Appropriate technologies for farmers in semi-arid West Africa*. H. Ohm and J.G. Nagy (Eds.). pp. 9-32. Purdue University Press, West Lafayette.

31. Odigboh, E.U. (1999). Machines for Crop Production in B.A. Stout and B. Cheze (eds.) CIGR.

32. Ofori C. S. (1993). Chapter 7: The challenge of tillage development in African agriculture. Soil Tillage in Africa: Needs and Challenges. FAO Corporate Document Repository, Economic and Social Development Department.

33. Orth, R. J., S. R., Marion, S. Granger, and M. Traber (2009). Evaluation of a mechanical seed planter for transplanting Zostera marina (eelgrass) seeds. *Aquatic Botany*, 90(2): 204-208.

34. Onwude D.I., R. Abdulstter, C. Gomes and N. Hashim (2016). Mechanization of largescale agricultural fields in developing countriesa review, *Journal of the Science of Food and Agriculture* 96: 3969–3976.

35. Pawar, C.S., N. A. Shirsat, and S.V. Pathak (2008). Performance evaluation of combine harvester and combination of self-propelled vertical conveyor reaper with thresher for wheat harvesting. *Agriculture Update* 3 (1 and 2): 123-126.

36. Qadir, M., E. Quillérou, V. Nangia, G. Murtaza, M. Singh, R.J. Thomas, P. Drechsel and A.D. Noble (2014). Economics of salt-induced land degradation and restoration, *Natural Resources Forum*, 1-15, November.

37. Rathke, G. W., T. Behrens, and W. Diepenbrock (2006). Integrated nitrogen management strategies to improve seed yield, oil content and nitrogen efficiency of winter oilseed rape (Brassica napus L.): a review. *Agriculture, Ecosystems and Environment*, 117(2), 80-108.

38. Razzaq, A., B.C. Ahmad and C.B.A. Sabir (1992). A comparative study of partial vs complete mechanized harvesting and threshing of wheat. *Agri. Mech. in Asia, Africa and Latin America*. 23(1): 42-44.

39. Rusu T., P. Gus, I. Bogdan, P. Moraru, A. Pop, D. Clapa, and L. Pop (2009). Soil tillage conservation and its effect on erosion control, water management and carbon sequestration. *Geophys. Res. Abs.* 11: 1481.

40. Muhammad S, Mueen-u-Din, M. Ali, L. Ali, M. Q. Waqar, M. A. Ali, L. Khalid (2015). Grain Losses of Wheat as Affected by Different Harvesting and Threshing Techniques. *International Journal of Research in Agriculture and Forestry*. 16(2): 20-26.

41. Shamabadi, Z. (2012). Measurement the wheat losses in harvesting stage. *Int. J. Agri. Crop Sci.* 4(23): 1797-1802.

42. Sims B. and J. Kienzle (2016). Making mechanization accessible to smallholder farmers in Sub-Saharan Africa. *Environments*, 3, 11, 1-18.

43. Smedema, Lambert K., and Walter J. Ochs (1998). Needs and prospects for improved drainage in devel USDA. 1987. Farm Drainage in the United States, History, Status, and Prospects.

44. Smedema, Lambert K. (2000). Global drainage needs and challenges: The role of drainage in today's world. 8th International Drainage Workshop, New Delhi, India.

45. Tefera, T., S. Mugo, and Y. Beyene (2016). Developing and deploying insect resistant maize varieties to reduce pre-and post-harvest food losses in Africa. *Food Security*, 1-10.

46. UNEP (1992). World Atlas of Desertification. Edward Arnold, London, UK.

47. Unger, P.W., B. A. Stewart, J. F. Parr, and R. P. Singh (1991). Crop residue management and tillage methods for conserving soil and water in semi-arid regions. *Soil Tillage Res.*, 20: 219-204.

48. M. Veerangouda, Sushilendra, K. V. Prakash And M. Anantachar (2010). Performance evaluation of tractor operated combine harvester. *Karnataka Journal of Agriculture Sciences* 23(2): 282-285.

49. Waller P., and M. Yitayew (2016). Center Pivot Irrigation Systems. *Irrigation and Drainage Engineering*, p. 209-228.

50. World Bank (1991). National Drainage Project, Arab Republic of Egypt. Staff Appraisal Report, Report No. 9792-EGT, Washington DC, USA.

*Chapter 7*

# Application of Standardized Precipitation Index (SPI) Computation for Drought Monitoring and Management in Bukit Mierah Dam, Malyasia

*Nurazlina Mohd Zaid\*, Mohd Zaharifudin Muhamad Ali*
*and Mohamad Hasmiruddin Mohd Nasaruddin*

*\*Senior Assistant Director, Department of Drainage and Irrigation,*
*Ministry of Natural Resources and Environment, Kuala Lumpur*
*E-mail: nurazlina@water.gov.my*

## Abstract

Droughts are an occurrence in most countries in the world. Each country has developed its own systems to classify their severity and to monitor their spatial and temporal variation. In Malaysia Standardized Precipitation Index (SPI) has been in use for long. Application of this system was investigated on an intensive scale in one of the more drought prone areas in the country and results of the same are presented in this paper. These show that the threshold values of SPI of various categories of dryness stand in need of some improvement. The findings suggest scope of such an exercise for other parts of the country also.

*Keywords*: *Drought, Standardized Precipitation Index (SPI), Bukit Merah Dam.*

## Introduction

Drought is considered as part of regular cycle in climate almost everywhere in the world. It is a periodic occurrence of natural vulnerability described as having below normal precipitation over an extended period of time (Homdee *et al.*, 2016). Drought is considered as one of the least understood natural catastrophes in the world (Chang *et al.*, 2016). It is a challenge to define and distinguish the traits. Differences in spatial extent, duration and intensity lead to the complexity of drought where no two droughts can be alike (Kelley *et al.*, 2016). Defining drought is rather subjective matter depending on individual's or national perspectives. From a hydrologist point of view, drought means prolong episode of abnormal precipitation while an agriculturist would interpret drought where the crop yields get seriously undermined.

Drought is usefully classified in four categories: meteorological, hydrological, agricultural and socio-economic (Wilhite and Glantz, 1985). Wen *et al.* (2011) stated all droughts begin as meteorological droughts due to precipitation shortage for prolong period of time. Hydrological droughts occur as a result of insufficient for surface or groundwater flows. Agricultural droughts take place when shortage of soil moisture effect average crop growth. Socio-economic droughts happen as consequences to water supply shortage that lead to adverse impact to people's health and quality of life. Despite these consequences, it is impossible to prevent or avoid the occurrence of droughts. Thus, the way forward is a comprehensive and effective management of drought so as to lessen the adverse impacts of drought. The application of knowledge, techniques, skills and tools are necessary to effectively monitor and issue early warning to allow for better responses to drought preparedness and mitigation.

Selecting suitable drought indices according to the needs of the area is crucial in ensuring the efficiency of a drought monitoring system (Mendicino *et al.*, 2008). Over the years many drought indices have been developed to measure drought severity. In general, classification of drought indices are established according to the type of impacts they relate to (Zargar *et al.*, 2011).For example, Standardized Precipitation Index (SPI) is renowned for monitoring meteorological drought. Drought indices incorporate relevant data to reflect different events and conditions (Zargar *et al.*, 2014). Available information, spatial scales and type of drought determined are among the features to consider in selecting drought indices. No drought index is superior to others in every environment though certain indices are better suited according to particular regional applications (Ntale and Gan, 2003).

The purpose of this paper is to evaluate drought in Malaysia with Bukit Merah Dam (BMD) as the study area, which incidentally is known for repetitive exposure to drought episodes. Drought indice "Standardized Precipitation Index (SPI)" was selected to measure the drought severity due to its simplicity in requirement of precipitation record as input data.

## Study Area and Data

BMD is located in the state of Perak, Malaysia at latitude 5°.0329 and longitude 100°.6515. Located in the equatorial doldrum area, the climate features comprise uniform temperature, high humidity and abundant rainfall. It is very uncommon to have a whole day with completely clear sky even during periods of severe drought. The dam catchment area is 480km² with surface area of 40km². Location of the study area is shown in Figure 7.1. Construction of the dam began in 1902 and completed in 1906. It was built for agricultural irrigation and water supply

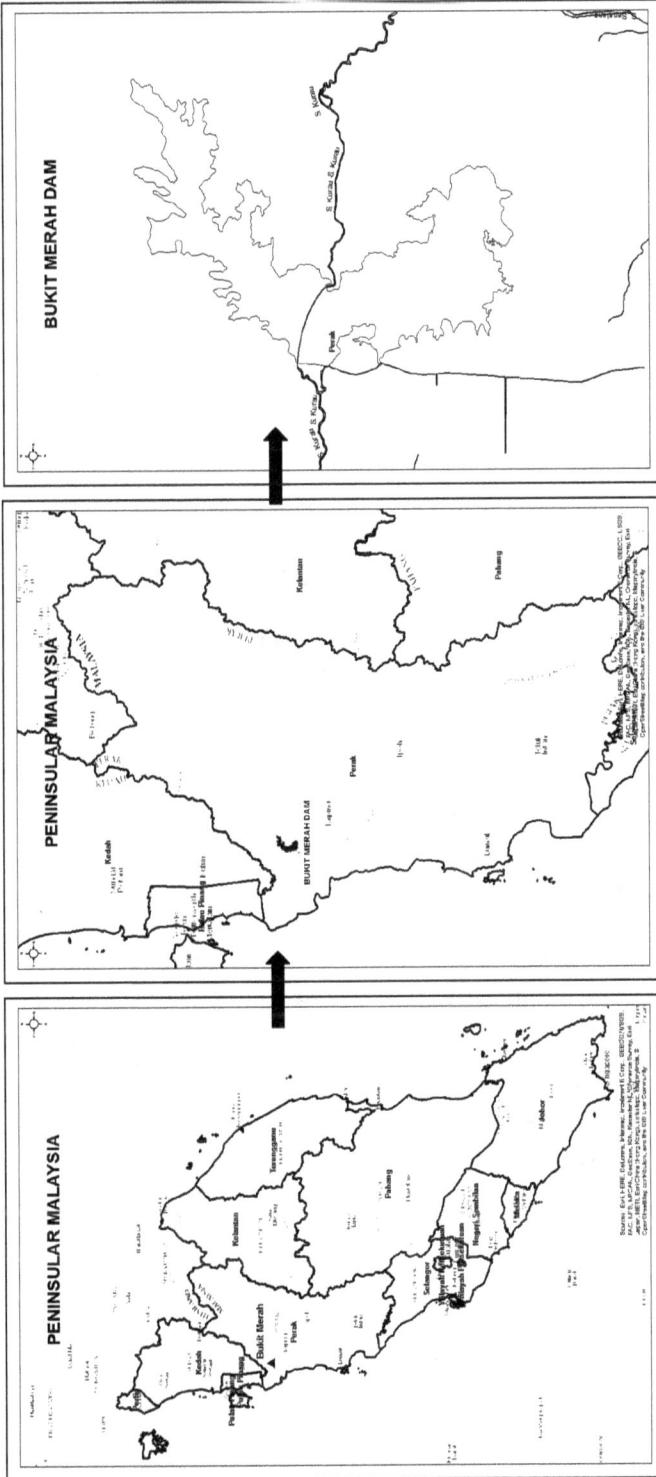

**Figure 7.1: Location of Study Area.**

purposes. Therefore, it is vital to understand the trend and transitions of droughts events from the commencement to termination in order to develop a dependable drought warning system.

The datasets used in this study is precipitation data from Bukit Merah Station. The station has an automatic system comprising data logger, tipping bucket rain, cables and software to accurately measure and record rainfall over long periods of time. The precipitation data of 64 years from 1953 to 2016 was retrieved from National Hydrology Network of Water Resources Management and Hydrology Division, Department of Irrigation and Drainage Malaysia (DID). The data was processed using Time Dependent Data (Tideda) software, version 4. Tideda is developed by National Institute of Water and Atmospheric Research (NiWA) for creating databases to store and analyse any time series data especially hydrological data.

## Methodology

Drought event in BMD was assessed using drought index, SPI, which was developed by T.B. McKee, N.J. Doesken, and J. Kleist in 1993. The SPI is a cumulative probability distribution of a long-term precipitation time series transformed into a normal distribution with a mean of zero and standard deviation of one. Theoretically, the SPI is a representation of the deviation of precipitation for a given location at a given time from 'normal conditions' (Venkataraman *et al.*, 2016). SPI computation is based solely on precipitation data, and does not take into consideration other climate variables such as losses through evapotranspiration. To achieve a good result, at least 30 years of precipitation data is needed. A distinctive feature of SPI lies in its versatility to monitor drought events on a variety of time scales from 1-month up to 72-months. Time scale is an important element in evaluating the impacts of drought in terms of commencement, intensity, duration and magnitude arid ending?? Looks in complete (B. McKee *et al.*, 1993). Furthermore, SPI has the ability to distinguish various types of droughts as certain systems and regions react to drought conditions at very different time scales (Vicente-Serrano *et al.*, 2012).

Calculation of SPI involves a long-term of precipitation data accumulated over selected time scale. The precipitation data is then fitted to a probability distribution before it is transformed into a normal distribution. A specific amount of precipitation for a specified time scale will be identified to a specific SPI value consistent with its probability (Lovino *et al.*, 2014). Positive SPI values indicate wet condition, while negative values indicate dry condition as illustrated in Table 7.1.

Drought disaster in Malaysia is managed according to the standard operating procedure (SOP) laid by the National Security Council of Malaysia (NSC). The SOP, released in 2011, was developed following several drought events in Malaysia particularly in the period 1992 to 1998. Issuance of drought early warning in Malaysia falls under the responsibility of Malaysian Meteorological Department (MMD). The release of warning is according to the SOP's levels of warning as illustrated in Table 7.2. The criteria to determine the levels of warning are based on rainfall deficit and SPI.

**Table 7.1: Wetness/Dryness Categories of Standardized Precipitation Index (SPI)**

| SPI Value | Wetness/Dryness Category |
|---|---|
| + 2 to more | Extremely wet |
| 1.5 to 1.99 | Very wet |
| 1.0 to 1.49 | Moderately wet |
| -0.99 to 0.99 | Near normal |
| -1.0 to -1.49 | Moderately dry |
| -1.5 to -1.99 | Severely dry |
| -2 to less Extremely | Extremely dry |

**Table 7.2: Issuance of Drought Early Warning in Malaysia**

| Level | Rainfall Deficit |
|---|---|
| **Alert** | ★ Total 3 consecutive months rainfall deficit >35 per cent |
| | ★ Current month SPI index < -1.5 |
| | or |
| | ★ Total 6 consecutive months rainfall deficit >35 per cent |
| | ★ Current month SPI index < -1.5 |
| Warning | ★ Total 3 and 6 consecutive months rainfall deficit >35 per cent |
| | ★ 3 months SPI index < -1.5 |
| | ★ Alert level declared |
| Emergency | ★ Total 3 and 6 consecutive months rainfall deficit >35 per cent |
| | ★ 3 months SPI index < -2.5 |
| | ★ Warning level declared |
| Termination | ★ SPI index positive value |
| | and/or |
| | ★ Current month rainfall above normal |

## Result And Discussion

Observed annual total precipitation for the historical period for DBM is presented in Figure 7.2. The results from computations of SPI for 1-month, 3-months, 6-months, 9-months and 12-months time scales are presented in Figures 7.3a-b. It can be seen that the SPI derived from precipitation for the observed dataset generally fall within the ensemble range. The frequency, duration and intensity of drought at BMD vary according to the time scales.

Analysis on the frequency of severely dry and extremely dry events based on 1-month, 3-months, 6-months, 9-months and 12-months time scale is shown in Figures 7.4a-b.The total number of severely dry and extremely dry events recorded from 1953 to 2016 is 252 events. Of all the drought events, severely dry constituted 63 per cent while extremely dry formed the remaining 37 per cent. The lowest value of SPI recorded has been -3.85, -3.27, -3.11, -2.57 and -2.79 at 1-month,

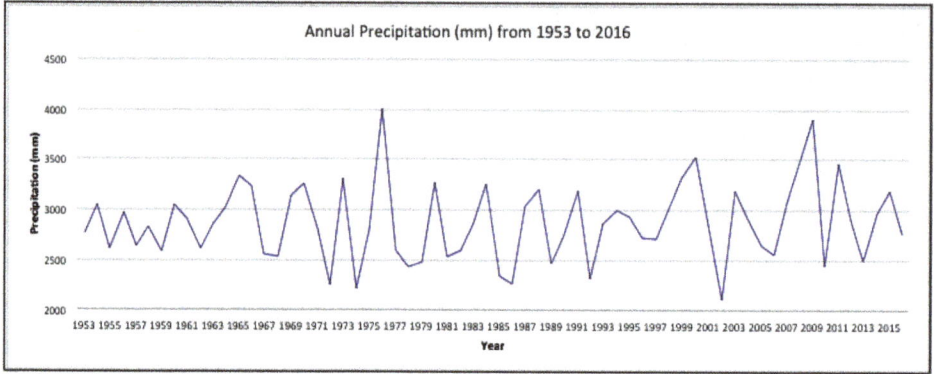

**Figure 7.2: Observed Precipitation for BMD.**

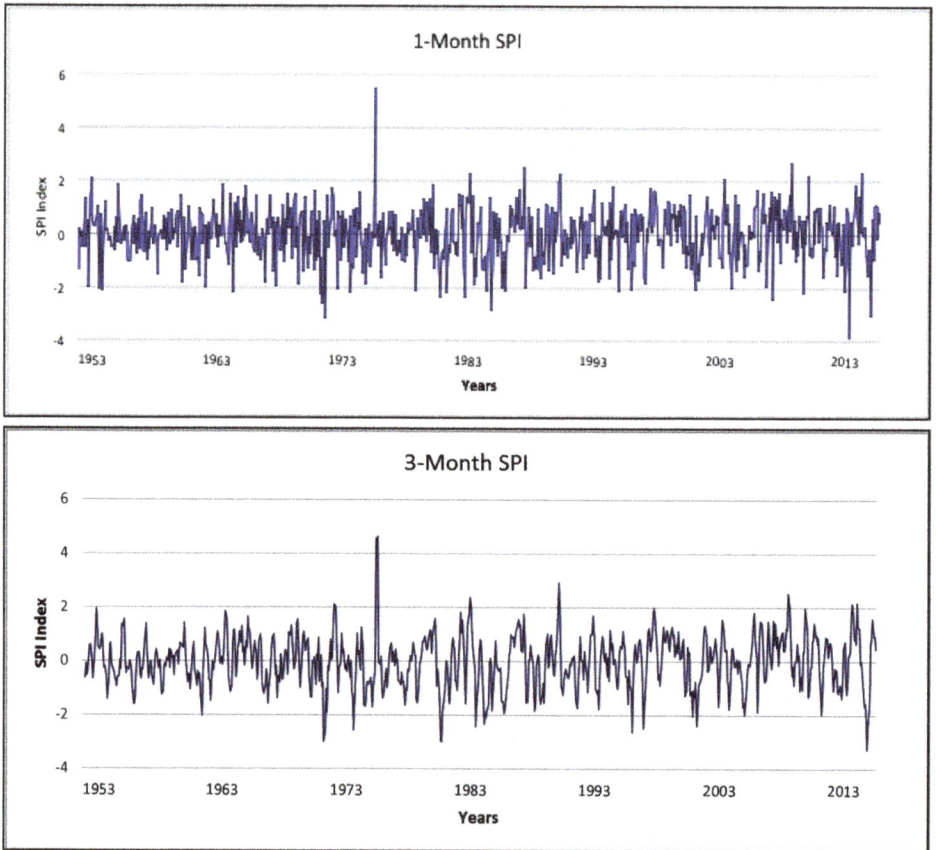

**Figure 7.3a: 1-month, 3-months Time Scales.**

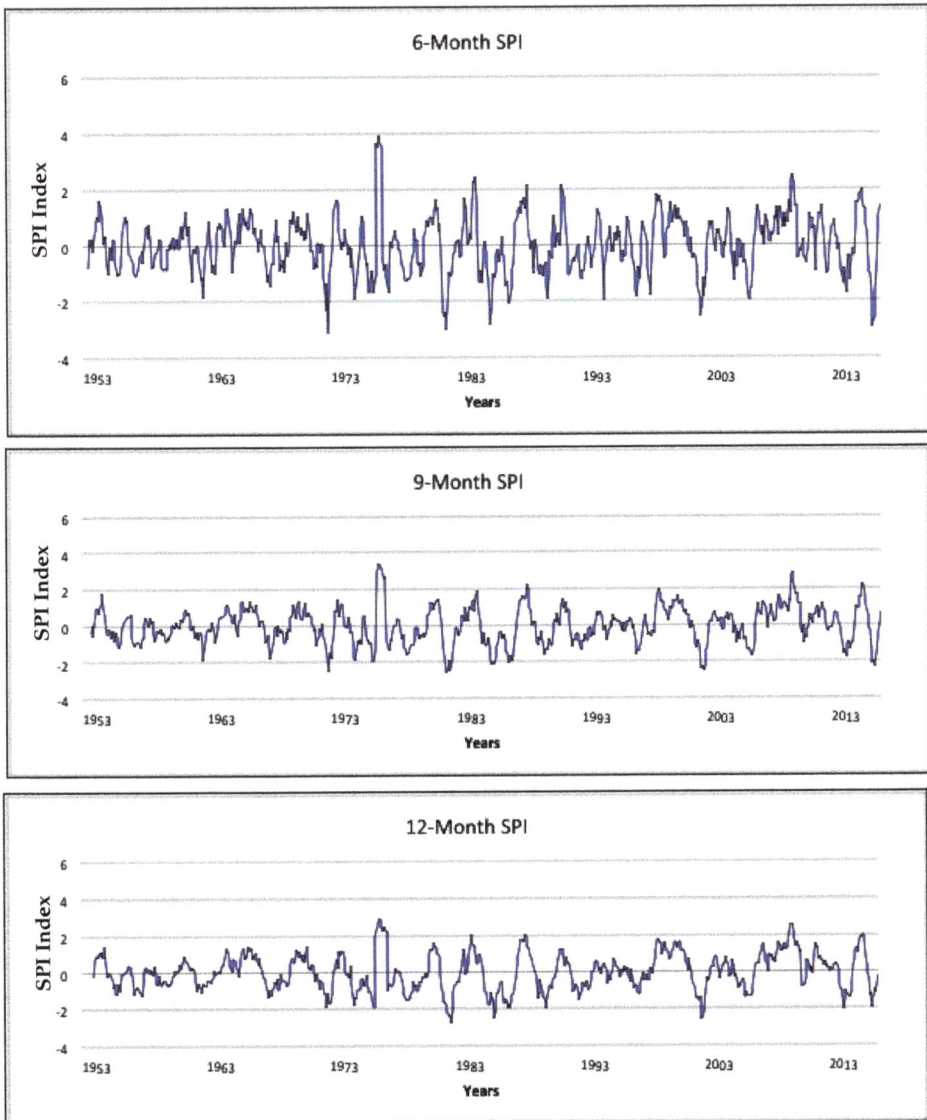

Figure 7.3b: 6-months, 9-months and 12-months Time Scales.

3-months, 6-months, 9-months and 12-months time scales respectively. SPI values for short-term time scales of 1-month and 3-months are dynamic. The values of SPI fluctuate due to the swift recovery of drought. However, the SPI can still be used for analysing if rainfall fluctuations is taken into consideration (Norouzi *et al.*, 2012). Thus, monitoring of drought events for BMD in this paper is focussed on longer time scales of 6-months, 9-months and 12-months.

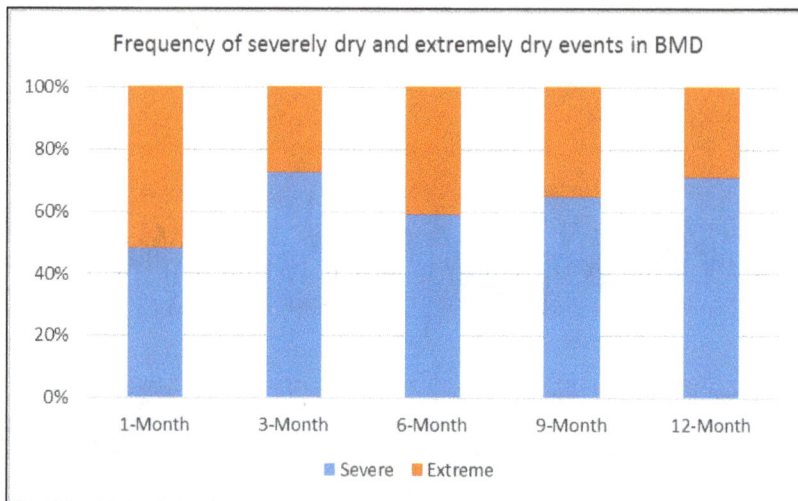

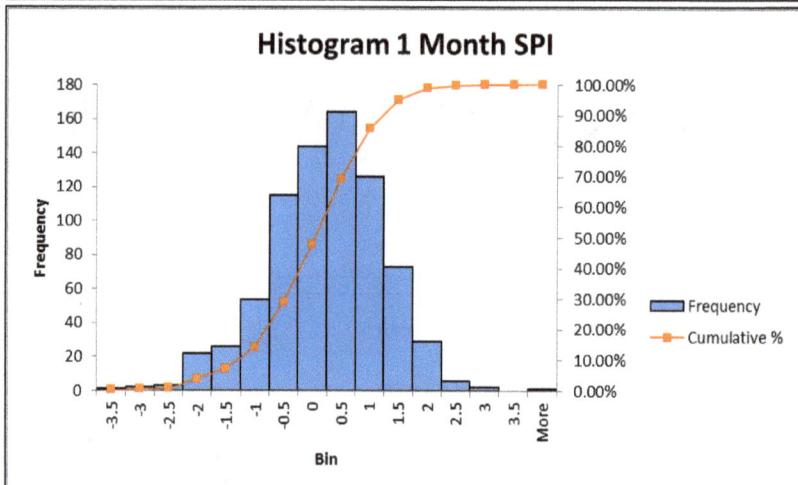

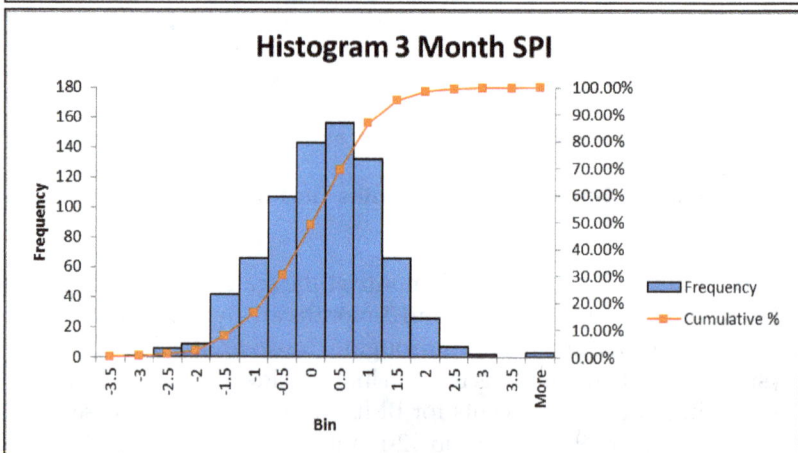

**Figure 7.4a: Frequency of Drought Events in BMD.**

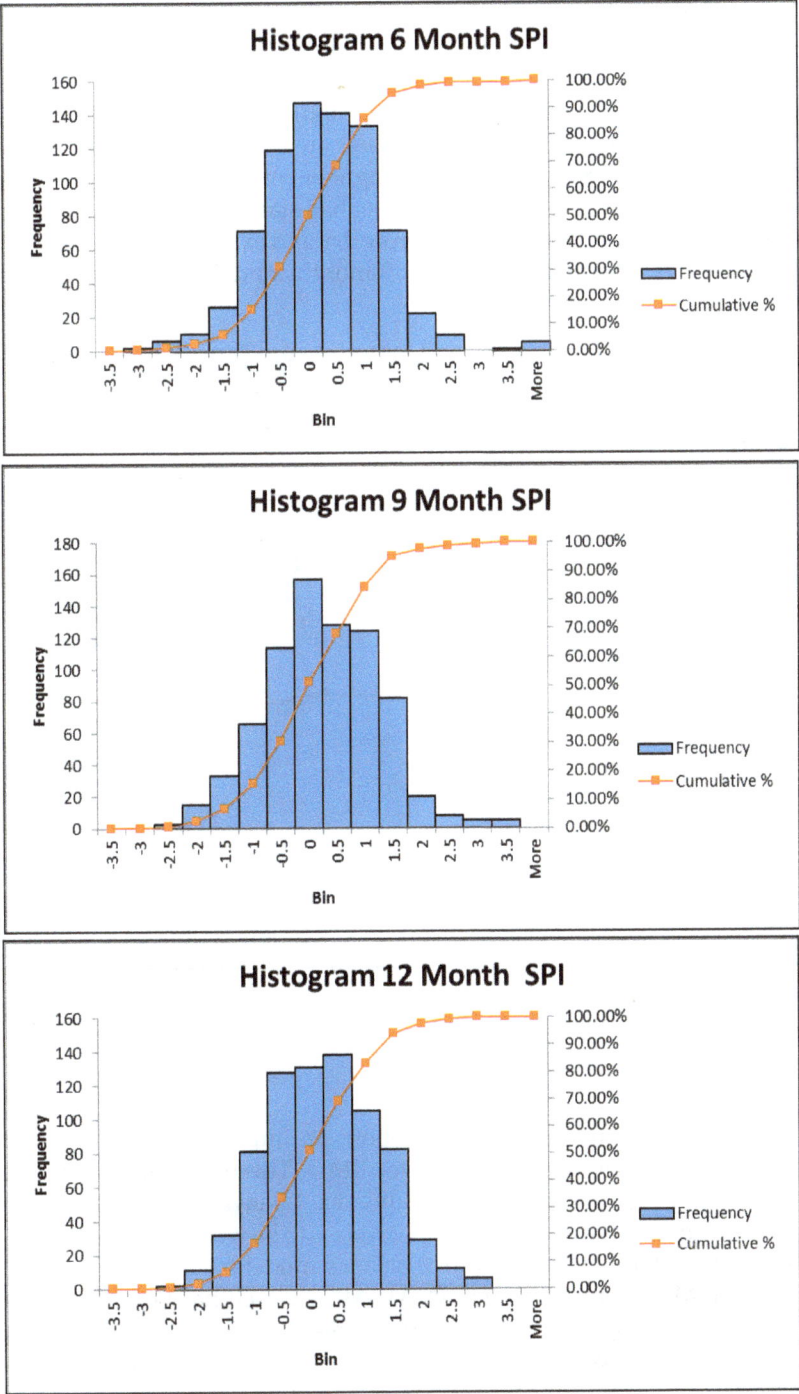

Figure 7.4b: Frequency of Drought Events in BMD.

The existing SPI does not differentiate the values of drought severity in term of time scales. The values are uniform regardless of the time scales. However, due to unique climate conditions in BMD, additional analysis was made to suit the condition of study area. Analysis and findings indicate the necessity for applying a new threshold. The proposed threshold values are based on recent drought events in BMD in 2014 and 2016. During both drought events, the study area experienced serious water crisis where dam and river water had started drying up. Determination of proposed SPI threshold values for severely dry and extremely dry events in BMD, based on 1-month, 3-months, 6-months, 9-months and 12-months time scales, is shown in Figure 7.5.

**Table 7.3: Proposed Severely Dry and Extremely Dry Threshold Values for BMD**

|                  | 1-month | 3-months | 6-months | 9-months | 12-months |
|------------------|---------|----------|----------|----------|-----------|
| **Severely dry**   | -1.65   | 1.55     | -1.2     | -1.30    | -0.95     |
| **Extremely dry**  | -3.00   | -2.80    | -2.0     | -1.75    | -1.55     |

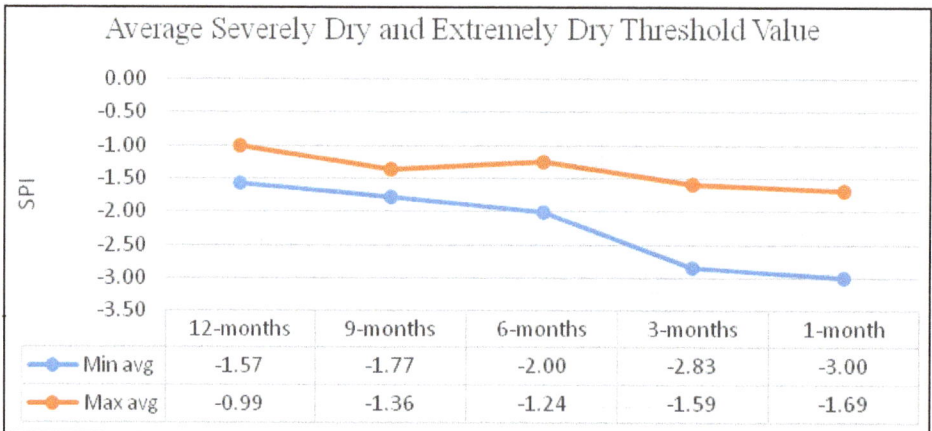

**Figure 7.5: Proposed Threshold Values for Severely Dry and Extremely Dry in BMD.**

## Conclusions

The objective of this paper is to propose an improved threshold values of SPI for severely dry and extremely dry for BMD. These values will be validated in future drought monitoring for the study area. Likewise, analysis on threshold values for other drought prone areas in Malaysia is to be worked out also. In keeping with the newly initiated water resource balance model for the country, known as Water Balance Management System (NAWABS), which has been formulated for the purpose of securing and sustaining water resources in Malaysia, new tailored threshold values for BMD and other areas hold promise for incorporation in the drought monitoring element of the model.

# References

1. Wilhite, Donald and H. Glantz, Michael (1985). "Understanding the Drought Phenomenon: Role of Definitions", *Water International* 10: 3, pp. 111–120.

2. McKee, Thomas; J. Doesken, Nolan and Kleist, John (1993). "The Relationship of Drought Frequency and Duration to Time Scales", Eighth Conference on Applied Climatology, 17-22 January 1993, Anaheim, California.

3. Homdee, Tipaporn; Pongput, Kobkiat and Kanae, Shinjiro (2016). "A Comparative Performance Analysis of Three Standardized Climatic Drought Indices in the Chi River Basin, Thailand", *Agriculture and Natural Resources* 50, pp. 211-219.

4. Jianxia Chang, Yunyun Li, Yimin Wang, Meng Yuan (2016). "Copula-based Drought Risk Assessment Combined with an Integrated Index in the Wei River Basin, China",Journal of Hydrology 540, pp. 824–834.

5. Kelley, Windy K.; Scasta, John Derek and Derner, Justin D. (2016). "Advancing Knowledge for Proactive Drought Planning and Enhancing Adaptive Management for Drought on Rangelands: Introduction to a Special Issue", Rangelands 38, pp. 159-161.

6. Lovinoa, Miguel; O. Garcíab, Norberto and Baethgen, Walter (2014)., "Spatiotemporal Analysis of Extreme Precipitation Events in the Northeast Region of Argentina (NEA)", Journal of Hydrology: Regional Studies 2, pp. 140–158.

7. M. Vicente-Serrano, Sergio; Beguerý´a,Santiago; Lorenzo-Lacruz, Jorge; Camarero, Jesu´ s Julio; I. Lo´pez-Moreno, Juan; Azorin-Molina, Cesar; Revuelto, Jesu´ s; Mora´n-Tejeda, Enrique and Sanchez-Lorenzo, Arturo (2012). "Performance of Drought Indices for Ecological, Agricultural, and Hydrological Applications", Earth Interactions 16.

8. Mendicino, Giuseppe; Senatore, Alfonso and Versace, Pasquale (2008). "A Groundwater Resource Index (GRI) for Drought Monitoring and Forecasting in a Mediterranean Climate", *Journal of Hydrology* 357, pp. 282– 302.

9. Norouzi, Akbar; Nohegar, Ahmad and Ghorbani, Ardavan (2012). "Comparison of the Suitability of Standardized Precipitation Index (SPI) and Aggregated Drought Index (ADI) in Minab Watershed (Hormozgan Province/South of Iran)", *African Journal of Agricultural Research* 7(44), pp. 5905-5911.

10. Ntale, Henry K. and Gan, Thian Yew Gan (2003). "Drought Indices and Their Application to East Africa", *International Journal of Climatology* 23, pp. 1335-1357.

11. Venkataraman, Kartik;Tummuri, Spandana;Medina, Aldo andPerry, Jordan (2016). "21st Century Drought Outlook for Major Climate Divisions of Texas Based on CMIP5 Multimodel Ensemble: Implications for Water Resource Management", *Journal of Hydrology* 534, pp. 300–316.

12. Wen, Li; Rogers, Kerrylee; Ling, Joanne and Saintilan, Neil (2011). "The Impacts of River Regulation and Water Diversion on the Hydrological Drought Characteristics in the Lower Murrumbidgee River, Australia", *Journal of Hydrology* 405, pp. 382–391.

13. Zargar, Amin; Sadiq, Rehan; Naser, Bahman and I. Khan, Faisal (2011). "*A Review of Drought Indices*", www.nrcresearchpress.com/er on 13 September 2011.

## Chapter 8

# Forestry Service Strategic Plan for Enhancing the Tree Cover of Mauritius 2016-2020

*Jhumka Zayd*

*Assistant Conservator of Forests,*
*Ministry of Agro Industry,*
*Food Security/Forestry Service, Phoenix, Mauritius*
*E-mail: zjhumka@govmu.org, zayd.jhumka@gmail.com*

## Abstract

Before the arrival of the first settlers around 420 years ago, Mauritius was totally covered with dense forests. After its human colonization, Mauritius experienced indiscriminate clearing of forest land for timber exploitation, agriculture, settlement and infrastructural development. In parallel, several invasive alien species which are still contributing to forest degradation have- been introduced. Large extent of native forests was also cleared and replaced by economically fast growing species such as Pinus elliottii and Eucalyptus tereticornis. In 2015, the extent of forest cover in Mauritius was estimated to be around 47, 100 ha, representing about 25 per cent of the total land area. However, the area of good quality native forest is estimated to cover only approximately 2 per cent of the island.

In recent years the Government has recognized that the forests and trees outside forests areas play a vital role in the sustainable development of the country owing to their environmental, ecological, social and economic functions. Consequently, conservation, protection and development of the remaining forests through sustainable management has become priority objectives of the Government and are reflected in the National Forest Policy of Mauritius. Yet, Mauritius being a Small Island Developing State (SIDS) with limited land resources, there is still much pressure on forest lands for agricultural and infrastructural developments. To further strengthen its commitment to increase tree cover and forest protection throughout the island, the Ministry of Agro Industry and Food Security commissioned a strategic plan for Enhancing the Tree Cover of Mauritius (2016-2020).

This report outlines the goals and objectives of the strategic plan and presents the action/implementation plan. It also summarizes the progress after two years of implementation as well as the main constraints faced during the implementation process.

*Keywords: Tree cover, Forest conservation, Strategic planning for enhancing tree cover.*

## Introduction

The Island of Mauritius forms part of the Mascarene archipelagos located in the Western Indian Ocean. Its mild tropical climate, topography and geographical isolation over a few million years has resulted in the evolution of diverse and unique flora and fauna. Prior to its human colonization, Mauritius was totally covered with forests and offered distinctive habitats such as, a rather dry palm forest in the north, a semi-wet evergreen forest consisting mostly of *Diospyros tessellaria* (Black Ebony) in the lowland and wet forests in the uplands. Small heath and dwarf forests were also found in the uplands. The arrival of people in Mauritius marked the beginning of the transformation of the natural environment.

The Island was successively colonized by the Dutch, French and British until its independence in 1968. For more than four centuries, Mauritius suffered from indiscriminate clearing of forest land for timber exploitation, agriculture, settlement and infrastructural development. In parallel, several invasive alien species which are still contributing to forest degradation have been introduced. Large extent of native forests was also cleared and replaced by economically fast growing species such as pine and eucalyptus. During the Dutch period, the black ebony forests were exploited for their timber. Under the French and British rule, agriculture was a key sector to ensure food security, rural employment and economic growth in Mauritius. Forest was being cleared mostly for agriculture and settlements. However, after its independence, Mauritius has gradually shifted from an agriculture-based economy to one driven by the manufacturing and tertiary sector. However, native forest was still being replaced by economically important exotic species for timber and poles for the construction industry and firewood for household purposes. Forest land was also cleared for settlement and infrastructural development. During its short settlement history Mauritius lost approximately 95 per cent of the original native forest was decimated. Forty-two higher plants, one bat species one butterfly, five reptiles, fifteen birds and thirty-six snail species became extinct.

In 2015, extent of forest cover in Mauritius was estimated to be 47 100ha, representing about 25 per cent of the total land area. However, the area of good quality native forest is estimated to cover only approximately 2 per cent of the island. There are two types of forest ownership in Mauritius: public and private. Privately-owned forest areas were estimated to be some 25,000 ha compared to some 22,103 ha which are state-owned lands. All State Forest land are protected by the Forest and Reserves Act. In contrast, only about 6540 ha of the total private forest lands (including river and mountain reserves) are legally protected by the Act.

Limited land area and increase in population keep a constant strong pressure on the remaining forest lands especially on private land. Moreover, because of the rising value of land, private forest owners are more inclined to convert their forests lands to more profitable land uses such as housing, business development and deer ranching. Between 1990 and 2010, Mauritius went through a phase of rapid economic growth and approximately 10 000 ha of forest land has been converted to other uses. The sharpest decline in forest area occurred between the year 2003 and 2004. The forest lands were cleared mostly for infrastructural developments, *e.g.* built up areas, roads, agriculture and reservoirs. As from year 2004 onwards, deforestation has been minimal due to a more rigorous implementation of the sustainable forest management principles. Over the last decade, the Government has recognized that halting and reversing the trend of deforestation will not only help to curb the Country's $CO_2$ emission but also provide additional environmental benefits. Despite being small in area, the forests of Mauritius perform various environmental and ecological functions that far outweigh their direct economic aspect. The roles of forests in soil and water conservation, carbon sequestration, conservation of biodiversity and genetic resources, recreation and ecotourism are now recognized and valued. Consequently, conservation, protection and development of the remaining forests through sustainable management are priority objectives of the Government and are reflected in the National Forest Policy of Mauritius (2006). In fact, the forests of the Republic of Mauritius are now managed more for these environmental and ecological functions rather than for the production of timber. Consequently, timber exploitation is gradually being phased out. A lot of importance

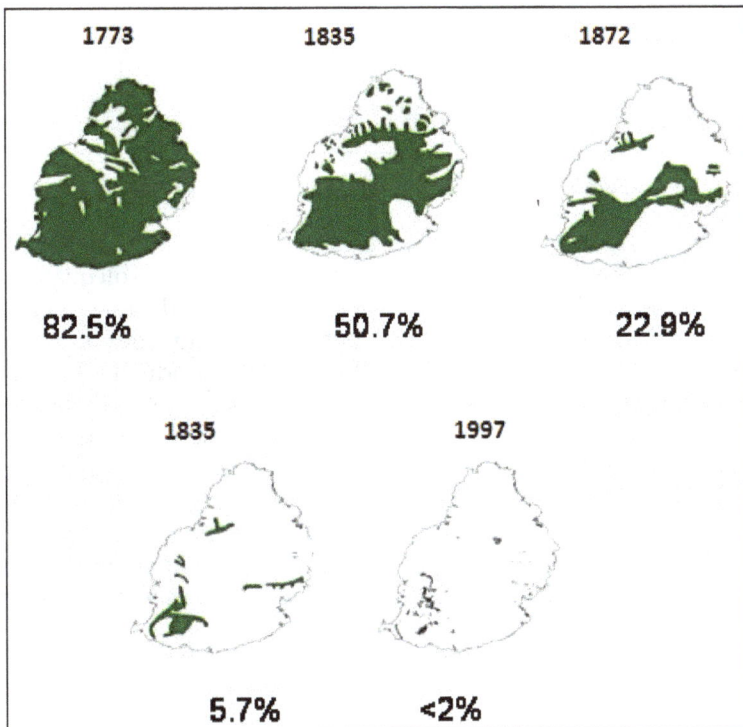

**Figure 8.1: Change in Native Forest Cover (*Source*: Mauritian Wildlife Foundation).**

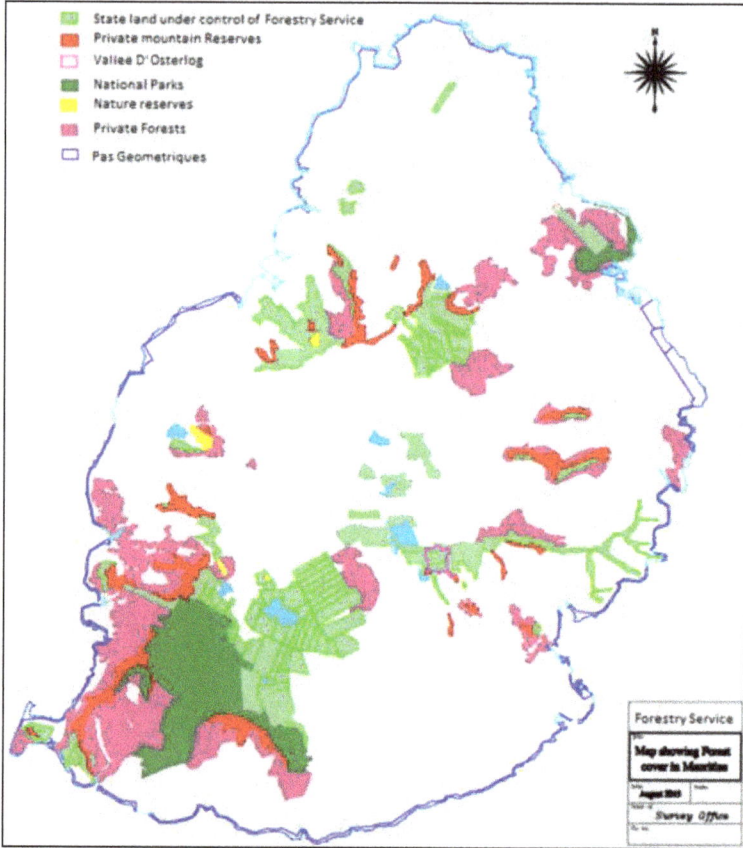

**Figure 8.2: Map Showing Forest Cover in Mauritius (Forestry Service, 2015).**

is also being given to trees outside forest areas. Urban trees offer opportunities for recreation, provide habitat for wildlife, improve air, water and soil quality, increase property value as well as absorb storm water run-off and prevent soil erosion. Despite the development of several laws, policies and strategies to protect forest, it was noticed that the tree cover in Mauritius was still gradually decreasing. The resulting vegetation loss and increase in hard surfaces is likely to cause economic, environment or human harm. To remediate the situation, the Ministry of Agro Industry and Food Security, requested the Forestry Service to prepare a strategic plan with the key objective of enhancing the tree cover of the Island. All aspects of the national strategic plan include the direction to implement a programme to enhance the tree cover in Mauritius. This report outline the strategic goals and action plan for the project and summarizes the progress after two years of implementation.

## Framework Strategy - Enhancing Tree Cover in Mauritius (2016-2020)

Preliminary works on the strategic plan started in February 2015 when the Forestry Service set up a team lead by the Conservator of Forests to prepare a nationwide tree planting strategy. A draft "Strategic Plan for Enhancing the Tree Cover of Mauritius 2015-2019" prepared in April 2017 and a consultative workshop was carried out on 23rd September 2015 to review and validate the draft strategic document. The workshop saw the participation of some 50 participants from various sectors involved in the forestry Sector, including, public institutions, parastatal bodies, NGOs and private land owners. The inputs and recommendations of the stakeholders were included into the draft "Strategic Plan for Enhancing the Tree Cover of Mauritius 2015-2019" which was submitted to the Ministry of Agro Industry and Food Security (MAIFS). The document was reviewed and approved by the MAIFS and was integrated as a chapter into the Ministry's "Strategic Plan (2016-2020) for the Food Crop, Livestock and Forestry Sectors." The implementation period of the strategies to enhance tree cover was accordingly changed to 2016-2020.

## Vision

To ensure a healthy green environment that will satisfy the needs and aspirations of present and future generations for environmental, social and economic benefits derived from trees in and outside forest areas.

## Mission

To enhance our tree cover and manage forests and trees outside forests with the participation and on behalf of the people of Mauritius.

## Strategic Goals

Four strategic objectives were derived from the mission statement in which effort should be focused for implementation of the strategic plan as follows:

### Strategic Goal 1: Tree Planting Programme to Increase Tree Cover throughout the Island

#### Specific Objectives

★ To promote enhancement of the carbon sink capacity and the environment by increasing forest and tree cover.

★ To increase native tree cover across the island.

★ To increase natural aesthetic values of the urban and rural landscapes.

★ To organize an effective campaign that would inform the targeted audience about importance of trees and forests, create a positive image, attempt to change the behaviour and influence social norm.

### Strategic Goal 2: To Create an Enabling Legal, Policy and Institutional Framework for an Economically Viable, Socially and Environmentally Sustainable Forest Sector

#### Specific Objectives

★ To enhance protection of trees on State lands and private lands.

⭐ To increase the number of trees to be replanted for every tree felled on state land, River Reserves, Mountain Reserves, Road reserves, Pas Geometriques, public beaches and public compounds.

⭐ To promote sustainable ecotourism activities on state forest land.

## Strategic Goal 3: To Ensure Sustainable Development, Management and Protection of Environmentally Sensitive Areas such as Watershed and Steep Slopes

### *Specific Objectives*

⭐ To increase planting of preferably native trees and other plants suitable for watershed protection around all lakes, reservoirs and in river reserves.

⭐ To increase planting of preferably native trees and other plants on steep slopes and coastal zones to reduce soil erosion.

⭐ To control environmental and agricultural development in water catchment areas and other ESAs.

⭐ To encourage private land owners to engage in protection and restoration of ESAs.

⭐ To raise awareness about the importance watersheds and wetlands.

## Strategic Goal 4: To Improve the Status of Biodiversity and Enhance the Benefits Derived from Environmental and Ecological Services

### *Specific Objectives*

⭐ To enhance biodiversity, wildlife habitat and ecosystem services.

⭐ To reduce direct pressures on biodiversity and promote sustainable land use.

⭐ To discourage the production, sale and use of known invasive alien species.

⭐ To eradicate or control invasive alien species in biodiversity rich areas.

⭐ To encourage private land owners to protect and restore native forest.

⭐ To promote sustainable ecotourism activities.

⭐ To raise awareness about importance of forests.

⭐ Eight broad action categories to achieve the goals and objectives were identified namely: (i) tree planting (ii) provision of incentives to private land owners and local authorities (iii) capacity building (iv) review of forest legislations (v) management of forests in ESAs (vi) Setting up of a coordination and liaison mechanism (vii) Research (viii) sensitization and awareness. These action categories and their strategic action sub-categories are overarching and stretch over the goal and objectives.

## Implementation Plan

### 1 Tree Planting

| Strategic Goals | Activities | Key Output | Output Indicators | Main Implementation Agencies | Estimated Budget (Rs M) |
|---|---|---|---|---|---|
| Goals 1,3,4 | Identify available planting space | New planting sites identified | Extent of planting space identified | Forestry Service (FS), Ministry of Local Government (Mo LG), DoE, NPCS, NGOs, etc. | 1 |
| Goals 1,3,4 | Set up a reforestation fund | Creation of a reforestation fund | Reforestation fund created | Ministry of AgroIndustry and Food Security (MAIFS),FS, State Law Office (SLO) | 1 |
| Goals 1,3 | Procurement of light mechanical equipment | Equipments to facilitate clearing and planting purchased | Tools/equipments/Irrigation network procured | MAIFS, FS | 4 |
| Goals 1,3,4 | Increase native species production capacity of forest nurseries and upgrading of equipment. | Extension of Nurseries. Logistics and equipment improved | Increase in forest nurseries production capacity | MAIFS, FS | 16 |
| Goals 1, 3, 4 | Upgrading Nurseries/ equipment for plant propagation | Nurseries propagation unit modernized | Plant propagation technologies improved in forest nurseries | MAIFS, FS | 2 |

### 2 Incentives to private land owners and local authorities

| Strategic Goals | Activities | Key Output | Output Indicators | Main Implementation Agencies | Estimated Budget (Rs M) |
|---|---|---|---|---|---|
| Goals 1, 3, 4 | Free issue of plants for planting along roadside and public areas | New plantations along roadsides and green spaces created | Number of plants issued and planted along roadsides | MAIFS, FS | 4.5 |
| Goals 1, 3, 4 | Free or subsidized seedlings to private forest land owners | New plantation established. Restored forests | Number of trees planted on private land | MAIFS, FS | 0.8 |
| Goals 1, 3, 4 | Provide private nurseries with free plant materials for propagation | Increase production and variety of species produced by private nurseries | Number Seeds/Cuttings/ Seedlings issued | MAIFS, FS | 0.4 |

| Strategic Goals | Activities | Key Output | Output Indicators | Main Implementation Agencies | Estimated Budget (Rs M) |
|---|---|---|---|---|---|
| Goals 1,3,4 | Developing incentives for private land owners to protect and restore pristine forests | Development of subsidies mechanism for private land owners | Subsidize herbicides/pesticides/fertilizers, counselling, free issue of native plants. Tax subsidies | MAIFS, FS. SLO, MoFED, Private Land Owners | 11 |
| **3  Capacity Building** | | | | | |
| Goals 1,3 | Train personnel in arboriculture and landscaping | Establishment of training and capacity building of forest officers and stakeholders in landscaping and arboriculture | Training module effectively implemented. Number of trained personnel | FS | 2.5 |
| Goals 3,4 | Capacity building in watershed management, invasive species management and conservation | Establishment of training and capacity building of forest officers and stakeholders in watershed management, invasive species management and conservation | Training module effectively implemented. Number of trained personnel | MAIFS, FS, NPCS, Water Resources Unit (WRU) | 7 |
| Goals 1,3,4 | Training on maintenance and management of new plantations | Competencies of staff on maintenance and management of new plantation strengthened | Number of trained staff | MAIFS, FS | 0.8 |
| Goals 3,4 | Provide training in conservation methods and ecosystem restoration | Competencies of staff in conservation and ecosystem restoration strengthened | Training module effectively implemented. Number of trained staff | FS, NPCS | 3 |
| Goals 1,3,4 | Train frontline staff education delivery methods, public speaking and use of multi media | Proficiencies of staff in delivery methods, public speaking and use of multimedia increased | Training module effectively implemented. Number of trained staff | FS, Ministry of Education (MoE), UoM | 4 |

| Strategic Goals | Activities | Key Output | Output Indicators | Main Implementation Agencies | Estimated Budget (Rs M) |
|---|---|---|---|---|---|
| **4** | **Review of Forest Legislations** | | | | |
| Goal 2 | Amend or prepare new the Forests and Reserves Act and Revise the Shooting and Fishing Lease Act | New and amended Forests and Reserves Act. | Forest and Reserves Act reviewed and passed through parliament | MAIFS/FS, State Law Office (SLO) | 9 |
| Goal 2 | Promote eco-tourism on State Forest Lands and impose a special levy to support reforestation through review of legislation | Amended legislation to support ecotourism project on leased state land | Legal amendments are documented and passed through parliament | MAIFS/FS, MoTourism (MoT), SLO, DoE | 2 |
| Goal 2 | Update the National Forest Policy | Reviewed forest policy | Updated and endorsed Forest Policy | FS | 1.5 |
| Goal 2 | Prepare a National Forest Action Program | Approved National Forest Action program | Approved National Forest Action plans are put in place | MAIFS, FS | 2.5 |
| Goal 2,3 | Control Infrastructural and agricultural development in water catchment areas through improved legislation | New or amended legislation for water catchment areas and ESAs | Legislation for protection of water catchment, watershed and other ESAs reviewed. | MAIFS/FS | 8 |
| **5** | **Management of plantation and natural forests in water catchment areas and other ESAs** | | | | |
| Goal 3,4 | Develop a forest rehabilitation model for biodiversity conservation in watershed and ESA areas | Rehabilitation model for biodiversity conservation in ESAs | Approved rehabilitation model for biodiversity conservation in watershed and ESAs put forward | MAIFS, FS, NPCS, NGOs, WRU, DoE | 2 |
| Goal 3 | Identifying and surveying main catchment areas | Main catchment areas surveyed | Extent of main catchment areas surveyed | MAIFS, FS, WRU, DoE, NPCS | 3 |
| Goal 3 | Survey erosion prone areas | Erosion prone areas surveyed | List of areas prone to erosion prepared. Vegetation surveyed in these areas. | MAIFS, FS, MoLG | 8 |

| Strategic Goals | Activities | Key Output | Output Indicators | Main Implementation Agencies | Estimated Budget (Rs M) |
|---|---|---|---|---|---|
| Goal 3 | Implementing a restocking programme in water catchment areas, steep slopes and other ESAs | Water catchment areas, Steep slopes and erosion prone areas replanted | Total area of Water catchment areas, Steep slopes and erosion prone areas rehabilitated/replanted | MAIFS, FS, WRU, NPCS | 18 |
| Goal 3,4 | Control of invasive species in ESAs | Restored/weeded ESAs | Extent of land weeded of alien invasive species in ESAs. | MAIFS, FS, NPCS, Vallee D'Osterlog, Private Land Owners, NGOs | 60 |
| Goal 3,4 | Gradual replacement of exotic forest plantation by native species | Increase Native forest areas | Area of exotic plantation replaced by native species | MAIFS, FS, DoE Private Land Owners, NGOs | 40 |
| **6  Setting up of coordination, communication and liaison mechanism** | | | | | |
| Goal 1,2,3,4 | Setting up a communication and coordination mechanism with other ministries and relevant stakeholders for concerted effort | Effective communication and Liaison Mechanism | Setting up of steering committee and sub-committees | MAIFS, FS | 2 |
| Goal 1,2,3,4 | Set up a coordination meeting of experienced technical Staff | Project evaluated, monitored and feedback mechanism provided | Coordinating body set up to monitor, evaluate and provide feedbacks on the project | MAIFS, FS, NPCS, DoE, MoHL | 0.2 |
| **7  Research** | | | | | |
| Goal 3,4 | Fostering research and cooperation with academic institutions and nongovernmental organisations | Top priority research areas determined and research studies commissioned | Workshop carried out; Bilateral meetings; Training Courses established | MAIFS, FS, NPCS, DoE Local Universities, FAREI, MSIRI, NGOs | 2.5 |
| Goal 3,4 | Conducting research on the best methods of eradicating/controlling invasive alien species | New methodological approaches for controlling invasive species put forward | Reports for each study. Pilot projects established | MAIFS, Forestry Service, NPCS, Universities, NGOs, MWF | 9 |

| Strategic Goals | Activities | Key Output | Output Indicators | Main Implementation Agencies | Estimated Budget (Rs M) |
|---|---|---|---|---|---|
| Goal 4 | Undertake phenological studies of endangered plants | Baseline data gathered | Creation of database. Reports for each study | FS,NPCS, NGOs, MWF | 8 |
| **8 Sensitization and awareness** | | | | | |
| Goal 1,3,4 | Generation of informative material and graphics on importance of forests and trees for dissemination through media (radio, newspaper, TV and Internet) | Public awareness on importance of trees, forest and biodiversity increased. | Materials distributed through media (Short films/ documentaries, newsletters, pamphlets, posters, painting and photography competition, *etc.*). | MAIFS, FS, NPCS | 5 |
| Goal 1,3,4 | Develop a pedagogical package of fun learning that is easily implementable in schools | Student awareness enhanced | Approved environmental and ecological programme in school. Books, games, quiz | MAIFS, FS, NPCS, MoE | 20 |
| Goal 1,3,4 | Propose special events within nature walks (events on International Day Celebration, art exhibitions, yoga sessions, trail expeditions *etc.*). Experiential learning. | Public awareness on importance of trees, forest and biodiversity enhanced. | Number of special events held (events on International Day Celebration, Open Days, *etc.*). Tree planting, weeding, photography events organized. | MAIFS, NPCS, FS, MoT | 4 |
| | **Total** | | | | 262.2 |

## Implementation Process

Due to its integration into the "Strategic Plan (2016-2020) for the Food Crop, Livestock and Forestry Sectors" the implementation of the "Strategic Plan for Enhancing the Tree Cover of Mauritius" started in January 2016 rather than 2015.

The implementation of the strategic plan Strategic Plan for Enhancing the Tree Cover of Mauritius is spearheaded by Ministry of Agro Industry and Food Security (MAIFS). The Forestry Service being the leading agency for the implementation of the strategic plan was tasked with the institutional coordination with key stakeholders, implementation, monitoring and evaluation for the implementation of the strategies to enhance tree cover. The Forestry Service reports on the implementation progress to the MAIFS on a regular basis. Several steering and advisory committees *e.g.* the RAMSAR committee, Environmental Impact Assessment Coordination Committee, National Invasive Alien Species Committee, Nature Reserves Board, National Parks Advisory Council also report directly to the MAIFS.

## Assessment of Progress after Two Years of Implementation of the Strategic Plan

### Tree Planting Operations and Incentives to Increase Tree Cover

The planting operations on roadsides and urban areas is coordinated by the Forestry service. The site selections for planting was carried out in consultation with all the stakeholders. The plants are provided by the Forestry Service and planting is carried out by the following institutions: Forestry Service, the Road Development Authority, Department of Environment, District Councils, Municipalities and NGOs.

During the period January 2016 – October 2017 more than 75 000 plants have been planted outside forest areas. This figure is however an underestimate because it does not take into account plants (around 45 000) which has been issued free of charge to Schools, socio-cultural organisations, NGOs and public. Other private initiatives for planting trees have also not been accounted.

A small decrease in the total area under forest cover (approximately 20 ha) was noted since the start of the implementation of the strategic plan in 2016. However, the overall stocking density of the state forest lands has been increased in some forest land, mostly due to the replanting of bare lands and filling of gaps within forest areas. More than 116 000 seedlings have been planted in bare lands/gaps within the forest areas, water catchment areas, steep slopes and other Environmentally Sensitive Areas, including some 21 129 natives species planted in previously weeded native forests. In the long run, this will increase the overall forest productivity. As an incentive to protect and restore native forests, an additional 6000 native plants were also issued free charge to private land owners to plant in their Mountain and River Reserves. Exemptions or reduction of taxes or leasing fees are also under consideration for private forests owners or lessees that interested who want to engage in conservation activities.

To meet the higher requirement for plants, the production capacity of the Forestry Service Nurseries had to be increased. So far two nurseries have already been upgraded. For example new overhead sprinkler systems were purchased and new shade houses were erected. Provision of free plant propagation materials to some private nurseries to boost the total plant production capacity was considered but has not been implemented.

Some new light mechanical equipments such as chainsaws, brush cutters, blowers, sprayers, *etc.* have been purchased to facilitate the clearing and planting process.

On the current trends, the original target of planting 100 000 trees annually over the 5-year duration of the strategic plant will be easily met.

## Review of Forest Legislation

Several gaps have been identified in the current forest legislations (*i.e.* the Forest and Reserves act of 1983, Amended 2003). The main gap being control of felling of trees on private lands. Legal reform can be a particularly lengthy process in Mauritius, particularly when new legislations is proposed to replaced existing laws. However, significant progress has been made in the forest code revision process.

The Forestry Service has obtained Technical Assistance from the Food and Agriculture Organization (FAO) for the project, "Support to Forest Code Revision and Institutional reform in Mauritius - TCP/MAR/3602". The project document was signed on 31 August 2016 between Mr Talla, the Regional Coordinator of FAO and the Honourable Minister of Agro Industry and Food Security, in the presence of the Director General of FAO. The project includes three components, namely: (i) design of the legal reform, (ii) design of the institutional reform and (iii) capacity building development. In connection with the project, three workshops have been carried out: (i) an inception workshop was held on 13th February 2017, (ii) a first consultative workshop was held on 22nd May 2017 and (iii) A second consultative workshop was held on 27th September 2017. A draft new forest legislation is expected to be ready July 2018.

It was proposed that the issue of control of agricultural and infrastructural development in water catchment areas be addressed during the revision of the Forest and Reserves Act.

The Shooting and fishing leases act (1966) has been amended in March 2017 to make provision for ecotourism activities on leased forest land. Regulations for the amended Act is currently under preparation.

As at 2017, almost no progress has been made toward the development of a national forest action program (NFAP). It was proposed that the NFAP be prepared after the new legislations and institutional reform has been completed.

## Rehabilitation of ESAs and Conservation of Biodiversity

The strategic document outlined various strategic actions for the protection and restoration of different types of Environmentally Sensitive Areas (ESA) such as wetlands, watershed, steep slopes and native forests. A number of decisions and initiatives have been taken and mechanisms established at the national levels to facilitate the protection and restoration of ESAs in the country.

The Protected Area Network Expansion Strategy (PANES) 2017-2026 was launched in May 2017. The project is co funded by GEF through UNDP. It makes provision for the expansion of the protected areas in Mauritius from 4.4 per cent to 16 per cent or more through the inclusion of land with high biodiversity and ecosystem value with the Protected Area Network (PAN). A thorough conservation mapping exercise was conducted to provide spatial guidance for the expansion. The PANES 2017-2026 also contains an in- depth review of the legal framework related to biodiversity conservation, outlines the institutional framework required for the implementation of the PANES and proposes the basis for biodiversity stewardship agreements for the incorporation of private land in the PAN.

A National Strategic Action Plan for the Conservation and sustainable use of Crop Wild Relatives (2016-2025) which aims to protect and restore the crop wild relatives (CWR) ecosystems for their conservation and sustainable used has been prepared and submitted to the Ministry of Agro Industry and Food Security for endorsement.

The Republic of Mauritius became an official partner of the Queen's Commonwealth Canopy (QCC) on 15 November 2016. The Forestry Service was mandated by the Ministry of Agro-Industry and Food Security to endorse 5 project under the umbrella of the Queen's Commonwealth Canopy. The project will include rehabilitation of 650 ha in 5 biodiversity rich areas (Black River Gorges national parks, Le Pouce Nature Reserves, Ilot Gabriel Nature Reserves, Vallée D'Osterlog Endemic garden, and Sir Seewoosagur Ramgoolam Botanical Garden).

As part of an ESA study, the country is reviewing updating the list and boundaries of ESAs across the country. Around 40 per cent of the ESAs have already been surveyed and demarcated.

With regards to the implementation of the restocking programme in water catchment area, steep slopes and other ESAs, preference is being given to native species because of their resilience and lower water consumption rates and other environmental and ecological benefits. Approximately 20 ha of bare lands and gaps within ESAs have replanted since January 2016.

Significant progress has also been made in the restoration of native forests. Since January 2016 to present. More than 150 ha of Native Forests located with National Parks, Nature Reserves, and catchment areas of reservoirs as been weeded of exotic species.

## Capacity Development

Training of staff at all levels is a cross cutting component of the strategic plan that influences the implementation of all the strategic objectives. During the preparation of the strategic plan the main areas of competency identified for strengthening included: (i) management of forest plantations and protected areas, (ii) monitoring and enforcement, (iii) public awareness and education and (iv) nature based tourism development (v) arboriculture and green Landscaping (vi) data management and information sharing.

Progress has been made in terms of capacity development initiatives that cut across the various strategic objectives and outcome areas. For example, the Government sponsored 20 forest officers for a Diploma Course in Forestry. The course has been specifically designed to meet the demand of the Forestry Service.

Short training programs were also run to strengthen capacity of Officers of the Forestry Service, National parks and Conservation Service, representatives of NGOs and private forest owners in diverse themes related to such forestry, endangered species management, habitat restoration, invasive s wildlife photography, sensitization and awareness, *etc.* These training programs were run under the Protected Area Network Project (GEF funded).

## Research

Apart from a few research project by students on ecology and conservation and propagation of native species, very little progress has been made towards implementation of the research components. The collaboration with academics and NGOs has not significantly improved since inception of the project implementation and research areas for the project has not been prioritised.

## Education Awareness Raising and outreach

The main difficulty with the any sensitization and awareness is to induce a change in behaviour and not restrict the campaign to information sharing only. In this perspective, experiential learning approach during the sensitization and awareness campaign was favoured. For instance, during specifically organised guided tours to Nature walks, the public were given the opportunity to plant native trees; Endemic gardens were created in schools with the participation of students; NGOs, youth clubs, and student participated in roadside and avenue planting; *etc.*

The communication and outreach strategy has also evolved from the more traditional dissemination of information through reports and articles to a more proactive engagement with the media, stakeholders and the public.

Sensitization and awareness campaigns have been carried out in 60 schools and 30 community centres. Guided tours to students are also regularly carried out by forest officers.

The following international events was celebrated in 2016 in 2017 to raise awareness: World Forest Day on 21st March, World Wetland Day on 2nd February 2017, World Biodiversity Day on 22nd May. These events were attended by students, representatives of various governmental organisations NGOs, private land owners and local inhabitants a NGOs. Information dissemination materials (pamphlets, posters, video clips, intervention on TV and Radio) are specifically prepared to marks these events. For experiential learning, tree planting events, creation of endemic gardens, free issue of endemic plants, photography competitions were organised.

## Constraints

In spite of key progress towards the achievement of certain goals, the implementation process faced several constraints. The main constraints are outlined below:

## Poor Coordination Process and Knowledge Management System

The coordination process for implementation, monitoring and evaluation of the activities has not yet been fully institutionalized and sustained. For example some stakeholders are still unclear on their roles in the plan implementation and monitoring. In addition, there was inadequate information sharing amongst institutions and stakeholders. There is no centralised database for baseline data and monitoring of the implementation process. An improve Knowledge management system is essential to leverages know-how across various institutions and to improve decision making, partnerships, increase transparency and overall results.

## Fragmented Institutional Arrangements

There is significant fragmentation in the responsibilities among various ministries and institutions in the implementation of the strategic plan including. *E.g.* Restoration of Native Forests is being carried out by the Forestry Service and National Parks and Conservation Service (both under the aegis of the MAIFS); there is much confusion regarding the management of Road side plantations, they are currently managed by the Road Development Authority, Forestry Service and Department of Environment (all three are under the aegis of different Ministries).

## Low Participation of Private Land Owners

Inability to stimulate sufficient private land owners to embark on forest restoration and protection activities. Due to limited land area, increase in population and rising value of land, private forest owners are more inclined to convert their forests lands to more profitable land uses such as housing, business development and deer ranching.

## Lack of Local Expertise

Some strategic actions have not been implemented due to a lack of local expertise. *E.g.* Development of a Rehabilitation model for biodiversity conservation in watershed areas.

## Lack of Base Line Data

There is significant baseline knowledge gaps regarding forests on private lands. Many wetlands have not been documented.

## Inadequate Human Resources

There are insufficient staff numbers, a shortage of qualified and experienced technical and administrative staff.

## Inadequate Financial Resources

So far the project is largely funded by the Government and additional funds will be required for sustained implementation of the strategic actions. One of activities proposed to mobilise funds (Setting up of a reforestation fund) has not been implemented. The project will have to develop a mechanism to mobilize funds from alternative source within the proposed time frame of the strategic plan to avoid any catastrophic financial constraints.

## Capacities Building Gaps

Some of the capacity building gaps that have been identified for technical and scientific areas and in core activity areas have not yet been implemented. *E.g.* Training in Landscaping and arboriculture.

## Conclusions

After two years of implementation of the action plan, significant progress towards the achievement of the strategic goals and objectives has been made:

- ★ Despite a small loss in forest acreage between 2015 and 2017, the overall tree cover outside forest areas and state forest land has improved.

- ★ Sensitization and awareness campaign emphasizing on experiential learning are being carried out regularly.

- ★ A major forest code revision and institutional reform is underway and will mostly like during the timeframe of the strategic plan.

- ★ Some progress has been made toward management practices in ESAs and some ESAs are being rehabilitated

- ★ Considerable effort has been put into restoration of native forests.

However, to fully achieve all the strategic goals and objectives the constraints outlined in the document will have to be addressed.

## References

1.  Bass, S.C. and Dalal-Clayton, D.B. (1995). Small Island States and Sustainable Development: Strategic Issues and Experience.

2.  Brouard, N.R. (1963). History of Woods and Forests in Mauritius. Forestry Service.

3.  Cheke, A. and Hume, J. (2008). Lost Land of the Dodo: An Ecological History of Mauritius, Réunion, and Rodrigues. *T and AD Poyser*, London.

4.  Cheke, A.S. (1987). An ecological history of the Mascarene Islands, with particular reference to extinctions and introductions of land vertebrates. In Studies of Mascarene Island Birds (ed A.W. Diamond), *Cambridge University Press*, London, 5-89.

5.  Convention on Biological Diversity (CBD) (2010). *Strategic Plan for Biodiversity 2011-2020: Further Information Related To The Technical Rationale For The Aichi Biodiversity Targets, Including Potential Indicators and Milestones.* CBD Document Reference COP/10/INF/12/Rev.1.

6.  Critical Ecosystem Partnership Fund (CEPF) (2014). *Ecosystem Profile: Hotspot of Madagascar and Indian Ocean Islands: Republic of Mauritius Synthesis Report – Preliminary Version February 2014.* Prepared by F. B. Vincent Florens for Biotope.

7.  Forestry Service (2000-2014). Annual reports, Ministry of Agro Industry and Food Security, Government of Mauritius.

8. Forestry Service (2015-2016) (in prep). Annual reports, Ministry of Agro Industry and Food Security, Government of Mauritius.

9. Global Forest Resources Assessment (FRA) (2015). Ministry of Agro Industry and Food security

10. Government of Mauritius (GoM) (2003). *Review of the National Development Strategy (NDS) Final Report: Volume 1: Development Strategy and Policies.* Ministry of Housing and Lands, Mauritius

11. Griffiths, O.L. and Florens, V.F. (2006). A field guide to the non-marine molluscs of the Mascarene Islands : (Mauritius, Rodrigues and Réunion) and the northern dependencies of Mauritius, *Bioculture Press.*

12. Gurib-Fakim A. (2003). Illustrated Guide to the Flora of Mauritius and the Indian Ocean Islands.

13. Hammond, D.S., Gond, V., Baider, C., Florens, F.B.V., Persand, S. and Laurance, S.G.W. (2015). Threats to environmentally sensitive areas from peri-urban expansion in Mauritius. *Environmental Conservation, 42(03),* pp.256-267.

14. Lalljee, B. and Facknath, S. (2008). A Study of the Historical and Present Day Changes in Land use Profile, and Their Driving Forces in Mauritius. Land Use: Reflection On Spatial Informatics Agriculture And Development, p.231.

15. Ministry of Agro Industry and Food Security of Mauritius (MoAIFS) (2015). *Fifth National Report on the Convention on Biological Diversity.* Port Louis, Mauritius.

16. Ministry of Environment (MoE) (2009). *Environmentally Sensitive Areas for Mauritius.* GIS mapping and reporting conducted by NWFS Consultancy, Portland, USA.

17. Ministry of Housing and Lands (2004). National Development Strategy. Government of Mauritius.

18. National Forest Policy for the Republic of Mauritius (2006). Ministry of Agro Industry and Fisheries, Forestry Service, Government of Mauritius.

19. NBSAP Ministry of Agro Industry (2006). National Biodiversity Strategy and Action Plan. Government of Mauritius.

20. NPSAP Ministry of Agro Industry (2017). National Biodiversity Strategy and Action Plan. Government of Mauritius.

21. Page, W. and D'Argent, G.A. (1997). A vegetation survey of Mauritius (Indian Ocean) to identify priority rainforest areas for conservation management, *IUCN/MWF report,* Mauritius.

22. PANES Ministry of Agro Industry (2017). Protected Area Network Expansion Strategy

23. Rouillard, G. and Gueho, J. (1999). Les plantes et leur histoires à Maurice.

24. Rughooputh, S., Jeetoo, C. and Daby, D. (2000). ICZM framework development: proposal for SIDS. *Mauritius: Faculty of Science, University of Mauritius.*

25. Vaughan, R.E. and Wiehe, P.O. (1937). Studies on the vegetation of Mauritius: I. A preliminary survey of the plant communities. *The Journal of Ecology*, pp.289-343.

26. Williams, J.R. (1989). Butterflies of Mauritius. 41 p.

*Chapter 9*

# Drought Assessment using Remote Sensing and GIS under Climate Change in the Mandalay Region of Myanmar

*Aung Myint*

*Assistant Lecturer,*
*Remote Sensing Department,*
*Mandalay Technological University, Myanmar*
*E-mail: aungmyintmg111@gmail.com*

## Abstract

In general, droughts have been classified into three categories in terms of impact which are: Meteorological drought, Hydrological drought and Agricultural drought.

Drought is the most complex but least understood of all natural hazards. It is broadly defined as "severe water shortage". Low rainfall and fall in agricultural production has mainly caused droughts. A drought impact constitutes losses of life, human suffering and damage to economy and environment. Droughts have been a recurring feature of the Myanmar climate therefore study of Historical droughts may help in the delineation of major areas facing drought risk and thereby management plans can be formulated by the government authorities to cope with the disastrous effects of this hazard.

Drought affects millions of people worldwide on an annual basis. Economically, it is the most devastating of all natural disasters. Drought is difficult to predict due to their slow onset. And in the aftermath, the long-term effects are often long lasting and widespread, making them difficult to recuperate from quickly. Some of these effects include loss of human and animal life, reduced crop and forest productivity, water scarcity and rationing, increased risk of fires, and damage to animal and fish habitats. Only a few countries have in place adequate drought mitigation strategies, instead relying on post-impact management strategies to deal with the effects of drought once the worst has transpired. One essential element to changing the discourse on drought is to incorporate it into our language and policy as an expected and natural part of climatic activity.

Drought conditions in Myanmar are the basis for further actions on drought management and for the development drought management policy at the country level. The dry zone, central area of Myanmar is the area vulnerable to drought as compared to other parts of the country. This dry zone area covers 67,700 km2 and 10 per cent of the total area. This area is characterized by low rainfall, intense heat and degraded soil conditions, affecting social and economic situations of the communities living in the region. There is about 35 per cent of the cultivable land in dry zone within the 3 regions (including 54 townships under 13 districts). The temperature is very high and hottest is in April and May. The precipitation in dry zone is controlled by the monsoon circulation system. Annual rainfall is less than 750 mm (national average precipitation is 2353 mm). In Myanmar, drought years were observed as 1972, 1979, 1982, 1983, 1986 and 1987. General description of drought conditions in 2010–2015 is crucial for drought studies and drought policy development. In this period the extreme temperature was recorded 47.2 °C on 14.5.2010 at Myinmu in dry zone area. The highest temperature was recorded at 20 stations during May. Inle Lake, which is the major tourist destination in Shan State of Myanmar, has been dried up in many parts. Water shortage was most severe in Ayeyarwaddy, Sagaing, Yangon, Mandalay and Bago Regions and Mon, Rakhine and Shan States in 2010. Most of the wells were dried up due to the depletion of undergroundwater supply because of late of Monsoon onset and so the scarcity of drinking water problems occurred in Myanmar. According to the Drought Annual Report of Department of Meteorology and Hydrology, the drought mostly occurred in dry zone area during Pre and Peak Monsoon period of 2010, the drought slightly occurred in Myanmar during 2011 and during 2012 and 2013, the severe and moderate drought occurred in dry zone area, some regions and states and mild drought occurred in some regions and states. The drought mainly impact to the agricultural fields, farmers, drinking water scarcity and also impact to social, economic, health, public, livestock and environment of Myanmar.

The NDVI and rainfall was found to be highly correlated (r=0.6) in water limiting areas. Apart from this, the highest NDVI-rainfall correlation associated with one-month time lager shows rainfall event induced vegetation growth in subsequent periods. The NDVI-rainfall correlation was found to be highly influenced by mean rainfall condition and vegetation types. Highest NDVI-rainfall correlation was obtained for vegetation types in rain fed crops, followed by irrigated crops and subsequently by forest with minimum correlation. It is therefore concluded that temporal variations of NDVI are closely linked with precipitation.

This drought risk area evaluation study involved the integration of Geographic Information Systems and Remote Sensing technology. Dry season data was acquired for 2 different years and was processed to detect vegetation condition change in response to drought. Physical and meteorological factors were analyzed and drought risk areas were identified based on the criteria of NDVI. NDVI change between a normal year (1995) and drought year (2015) was analyzed for each drought risk area. It was found that the value of the NDVI is lower in high drought risk areas, which justifies the modified criteria of NDVI.

*Keywords: Drought, Remote sensing and GIS, NDVI, Rainfall.*

# Introduction

Myanmar is located between 9° 55′ and 28° 15′north latitude and 92° 10′ and 101° 10′ east longitude. It is the westernmost country in Southeast Asia. Over 50 per cent of the eastern coastline of the Bay of Bengal and the Andaman Sea constitutes the western border. The country occupies a total land area of 676,577km$^2$ and is home to 57.5 million people in 14 states and regions as per 2008 estimate. The general topography of the country is high in the north and the west with north–south-oriented mountain ranges extending from the Himalayas. Present study area comprises the dry zone situated in the central area of Myanmar that is also vulnerable to drought as compared to other parts of the country. It makes use of remote sensing satellite data which is consistently available, cost effective and can be used to detect the onset of drought, its duration and magnitude. Moreover, an effort has been made to derive drought risk areas facing agricultural as well as meteorological drought using eight-year time series rainfall data and decadal SPOT satellite NDVI (Normalized Difference Vegetation Index) Output and employing a deviation of the current NDVI with the long term mean NDVI.

## Study Area

The high drought hazard area (Figure 9.1) lies in the central part of northern Myanmar. This is surrounded by the low hazard area on either side and by medium hazard area at the southern end.

The area of this high hazard zone is 67,700 square kilometers; about10 per cent of the total area of Myanmar. The region is characterized by low rainfall, intense

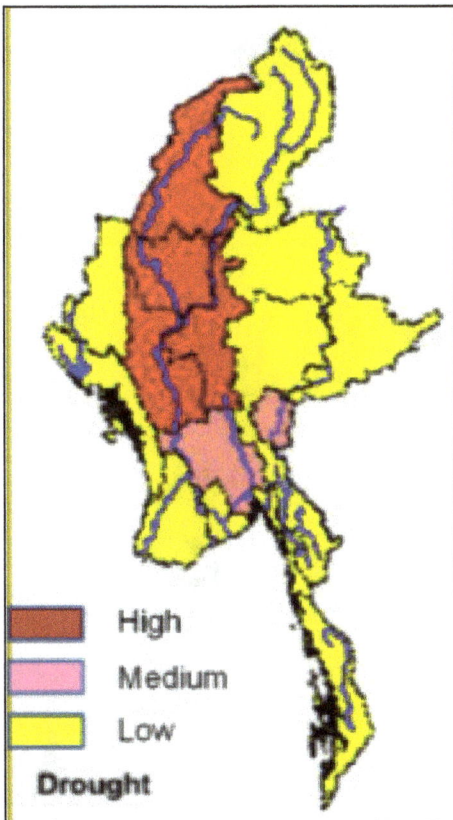

**Figure 9.1: Potential Drought Hazard Level of Myanmar.**

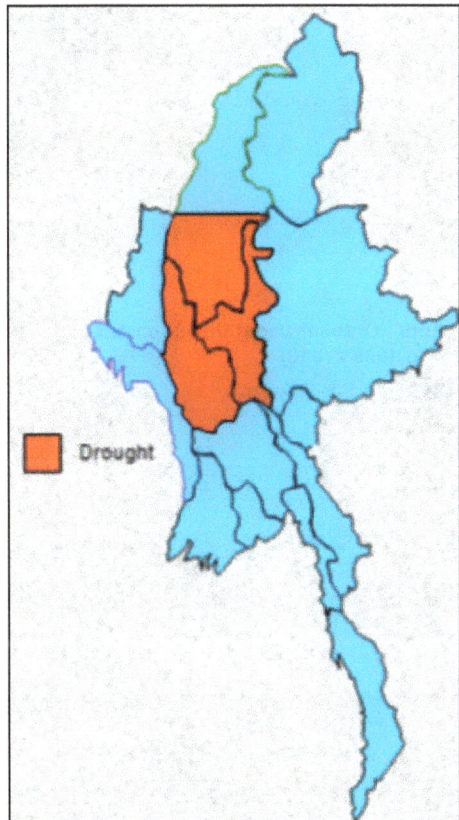

**Figure 9.2: Drought Vulnerable Area or Dry Zone Area.**

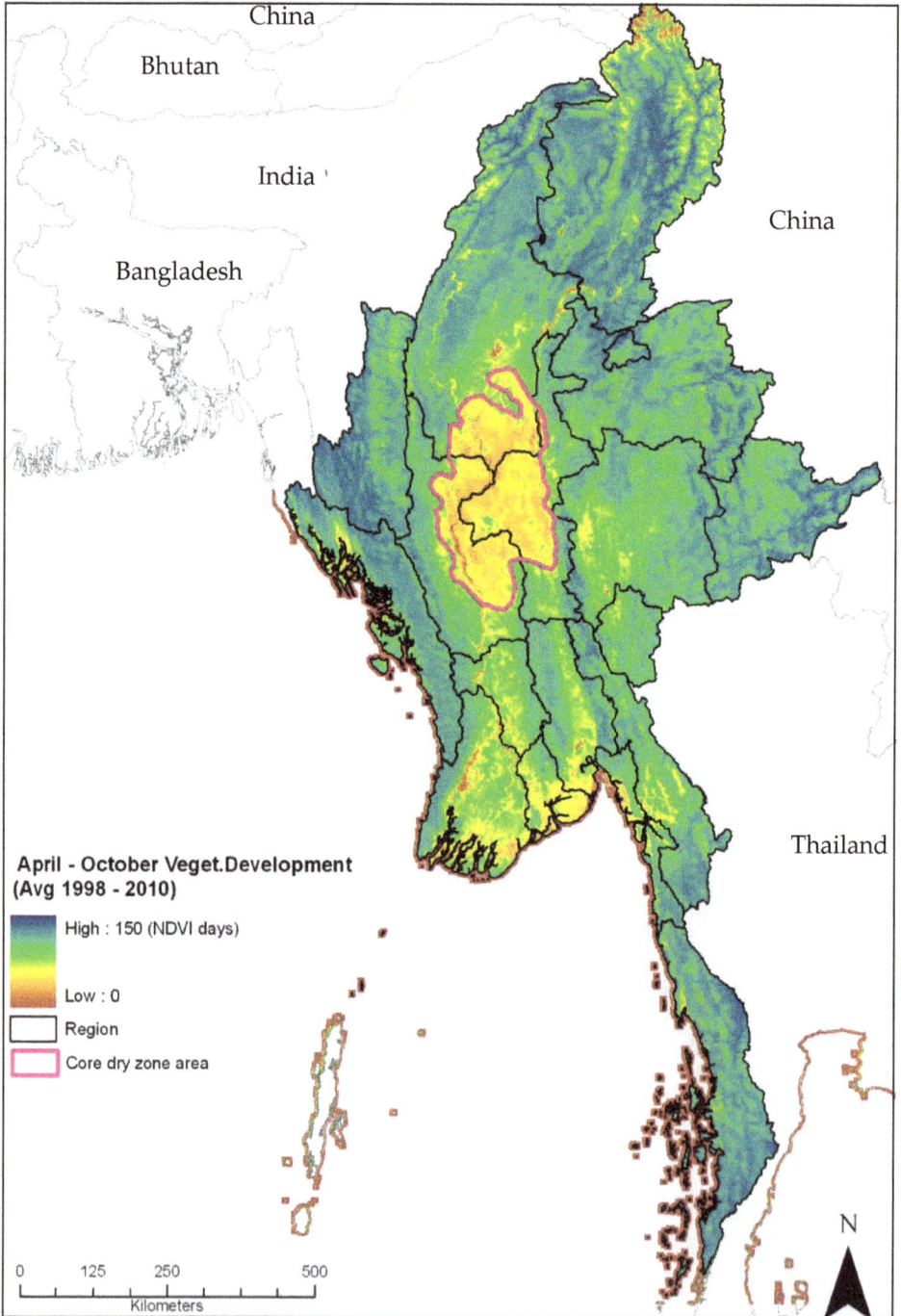

**Figure 9.3: Drought Vulnerable Area or Dry Zone Area and Average Vegetation Index in (1998-2010).**

heat and degraded soil conditions, affecting social and economic situations of the communities living in the region. Approximately 35 per cent of the cultivable land is located in dry zone. The drought vulnerable area also called dry zone area is shown in Figures 9.2 and 9.3. The temperature of the dry zone is very high and April and May are the hottest months. The precipitation in Dry Zone is controlled by the monsoon circulation system. The annual precipitation in dry zone is less than 750mm, while the national average precipitation is 2353.06 mm.

According to historical data, drought years were observed as 1972, 1979, 1982, 1983, 1986 and 1987. Analysis of drought indices of Myanmar (1951-2000) showing in Figure 9.4 which can be seen the drought indices of Myanmar are shown the rising trend. According to the Annual Drought Reports (2010-2013) prepared from Drought Monitoring Center of Department of Meteorology and Hydrology (DMH), Myanmar, the drought mostly occurred in dry zone area during Pre and Peak Monsoon period of 2010, the drought slightly occurred in Myanmar during 2011 and the severe drought occurred in dry zone area, the moderate drought also occurred in some region of dry zone area and also other regions and states and mild drought occurred in some regions and states during 2012 and 2013.

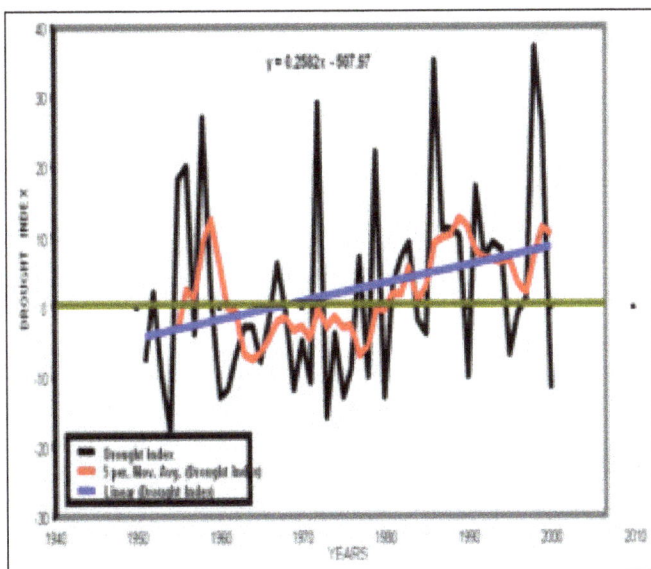

**Figure 9.4: Analysis of Drought Indices of Myanmar (1951-2000).**

The droughts in Myanmar mainly impact to the agricultural fields, farmers, drinking water scarcity and livestock's. According to the analysis of the annual lowest water level of the stations along the Ayeyarwady and Chindwin rivers in central Myanmar area the annual lowest water levels of these stations are showing the falling trends and also annual lowest water level recorded in 2010 at Monywa station and Mandalay station.

In Myanmar, the significant drought occurred during 2010. The extreme temperature also recorded 47.2 °C at Myinmu station in dry zone area on 14 May

2010. Myanmar was hit by a drought in 2010, which was the most severe in several decades. Temperature has been higher in this year than previous years in Myanmar and rain fall is late, causing severe shortage of water in many parts of Myanmar. In May, temperature has recorded the highest at 20 stations in Myanmar according to the observed data of DMH. In April, temperature has raised as high as 40 degree Celsius, according to the DMH's data observations. In some parts of Myanmar, temperature is as high as 43 degree Celsius. As a result, many streams and water reservoirs were dried up all over Myanmar. Inle Lake, which is the major tourist destination in Shan State of Myanmar, has been dried up in many parts. Water shortage is most severe in Ayeyarwaddy (Irrawaddy), Sagaing, Yangon (Rangoon), Mandalay and Bago Regions and Mon, Rakhine and Shan States. Most of the wells were dried up due to the depletion of undergroundwater supply because of late of Monsoon onset and so the scarcity of drinking water problems occurred in Myanmar. So the droughts impact to social, economic, health, public, livestock's and environment of Myanmar.

## Drought Monitoring and Early Warning Systems

Drought monitoring work, over the years, has been undertaken mainly by the Ministry of Agricultural and Irrigation. DMH has one center in Mandalay that devotes to drought monitoring and forecasting and this can be further improved. The cooperative efforts between various concerned agencies such as agricultural planning, irrigation, health, dry zone greening, forestry, National conservation for environmental affairs and livestock breeding will also be encouraged for drought management in Myanmar.

In Myanmar, there are 63 Meteorological Stations, 28 Hydrological Stations, 39 Meteorological and Hydrological Stations, 17 Agro-meteorological Stations, 8 aviation weather stations and 1 upper air station under the Department of Meteorology and Hydrology (DMH).

The early warning system is the main responsibility of DMH for disaster risk reduction in Myanmar. DMH issues the daily, dekad, monthly and seasonal weather and water level forecasts, news, warnings and bulletins for storms, floods, untimely rainfall and temperature *etc.* and also issues the Minimum Alert Water level and Bulletin for 7 stations in dry zone area during low flow period. DMH also issues the dekad agro-meteorological bulletins to support the agriculture. In these bulletins, the situation of soil moisture water balance, rainfall, temperature, relative humidity and evapo-transpiration of all regions and states of Myanmar are included. DMH also established the drought monitoring center at the upper Myanmar office (Mandalay office) locating in dry zone area in 2010. This center is now preparing and issuing the seasonal and annual drought reports based on the rainfall conditions. DMH cannot issue the warnings for drought management. So DMH needs to upgrade the drought monitoring center such as capacity building and also forecasting techniques *etc.* And also DMH is trying to upgrade the data observation networks, the forecast techniques, capacity building and the early warning system cooperating with international organizations.

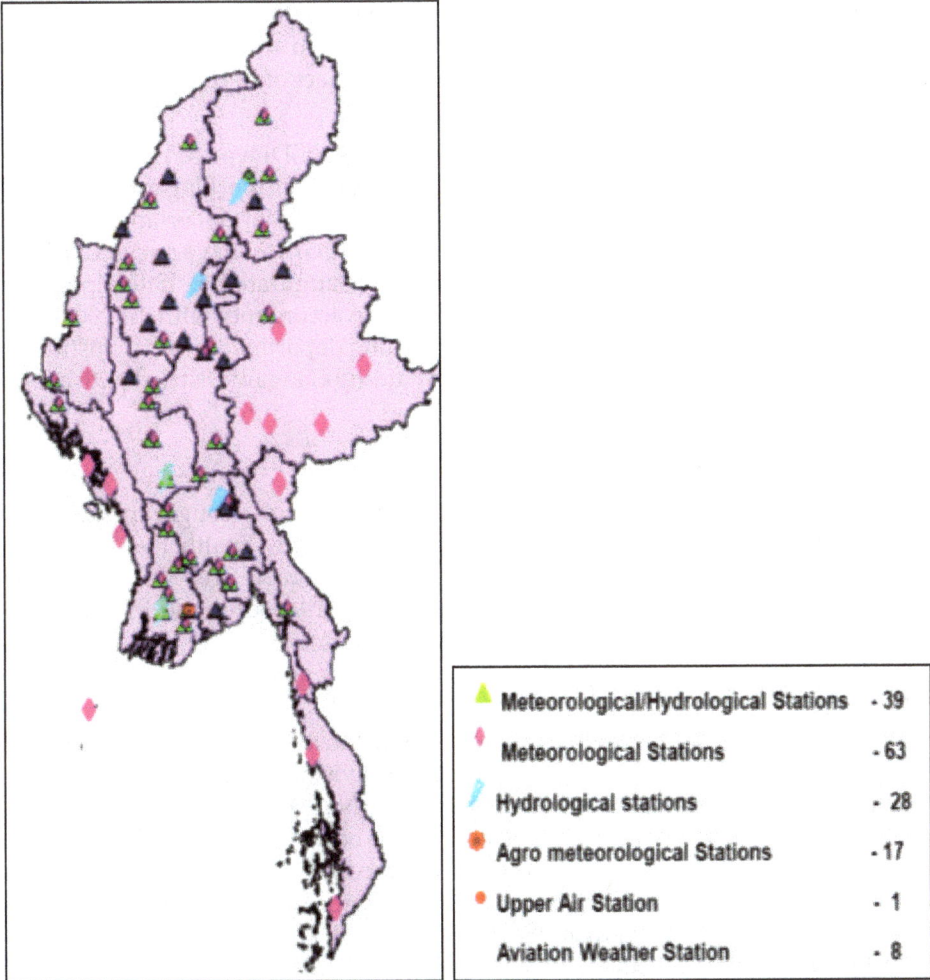

| | |
|---|---|
| ▲ Meteorological/Hydrological Stations | - 39 |
| ♦ Meteorological Stations | - 63 |
| / Hydrological stations | - 28 |
| ● Agro meteorological Stations | - 17 |
| ● Upper Air Station | - 1 |
| Aviation Weather Station | - 8 |

**Figure 9.5: Meteorological and Hydrological Observation Network (DMH).**

The other relevant department such as Forest department were carried out Dry zone rehabilitation activities since 1953, Agricultural and Rural Development Corporation (ARDC) was formed and development activities were carried out in central dry zone of Myanmar from 1953-54 to 1963-64 (10-year period). By the year 1963, the Forest Department (FD) succeeded responsibility of the task. Two 10-year working plans (1963-64 to 1972-73 and 1972-73 to 1981-82) were drawn up for the period of 20 years (from 1963-64 to 1981-82) and implemented for the development of central dry zone. At that time activities were mostly concentrated in Meikhtila forest district, particularly for reforestation of watershed areas, establishment of village-owned-forests and model forests in Mount Poppa watershed area.

After constitutional reform of FD in 1982 (withdrawal of district level administration) and after 1988 disturbances, starting from 1994-95 as a first 3-year pilot project, Special Region - Nine District Greening Project was adopted and carried out by FD. In 1995-96 the project had extended from nine districts to 13 districts. Watershed Mountain Greening Special Project of Myingyan district was also started in 1996-97.

In July 1997, Dry Zone Greening Department (DZGD) was constituted under Ministry of Environmental Conservation and Forestry. Its working covers central dry zone of Myanmar including 3 regions (Sagaing, Mandalay and Magway Regions), 13 districts and 57 townships, covering 21.557 million acres of dry land forests. The headquarters of the department was inaugurated in 18th September 1997 at Patheingyi Township, Mandalay division. In accordance with the 2000-2001 amendment, the working area of dry zone greening department was reconstituted as 3 regions, 12 districts and 54 townships (excluding Gangaw District) with a total coverage area of 20.17 million acres.

## Vulnerability Assessment

The vulnerable sectors of society and economy due to drought in Myanmar are agriculture and food production, drinking water supply, health, livestock's and fisheries, industry and environment. The largest vulnerable area is the dry zone area of Myanmar and the effected societies are farmers, people and livestock's in rural area.

## Emergency Relief and Drought Response

The Government of Myanmar has established institutional arrangement and has systems and procedures at National, State/Division, District, Township and sub-township levels for Disaster Management. The National Disaster Preparedness Central Committee under the Vice President (2) has been re-established in May 2013, the 22 members are included and the members are the ministers of (19) concerned ministries, prime ministers of regions and states, deputy minister of ministry of social welfare, relief and resettlement and director general of government office. National Disaster Preparedness Management Working Committee under the Minister of Ministry of Social welfare, Relief and Resettlement has also been re-established in May 2013. Under this, the 10 sub-committees are established. The members of National Disaster Preparedness Management Working Committee are the ministers and deputy ministers of concerned ministries, the ministers from regions and states, Chairmen of 10 sub-committees and director generals of concerned departments. In Myanmar, these two committees, the concerned departments and organizations are cooperating and working for disaster management in Myanmar. These disaster management committees and the Ministry of Social welfare, Relief and Resettlement are the main responsibility of emergency relief and response for disaster and also other concerned departments and organizations are cooperating. So these committees, concerned ministries and departments and organizations carry out the activities for relief and drought response. The emergency relief and drought response in Myanmar should be upgrade.

## Practices to Alleviate Drought Impacts

The practices to alleviate drought impacts in Myanmar are as follow:

1. DMH is issuing the daily, decade, monthly weather and water level forecasts, news, warnings and bulletins for storms, floods, untimely rainfall, temperature and minimum alert water level, agro-meteorological bulletins and seasonal and annual drought report. DMH likes to upgrade the forecasting and warning system for drought management.

2. Agriculture department is also doing in cooperation with international seed research centers for seeking and identifying drought resistant crops in Myanmar, conducting research on cultivation methods to be employed at the time of drought. The Ministry has been importing seeds that can survive with less dependence on water, and trying to nurse the crops and produce them at home.

3. Irrigation Department under the Ministry of Agriculture and Irrigation carry out the construction, repair and maintenance of dams, reservoirs and water supply facilities.

4. The Ministry of Agriculture and Irrigation has been implementing not only dams, reservoirs and the river water pumping project but also helping the people to build drinking water supply works. It does so by feeding water to water tanks from the dams and reservoirs, digging lakes and wells, installing water purifying systems and providing other technology.

5. Ministry of Environmental Conservation and forestry (MOECAF) has been implementing the afforestation and land rehabilitation in dry zone area by the projects.

6. The local governments, public and also NGOs are also implementing the digging lakes for getting the drinking water during drought period, rainwater storage and distributing the drinking water during water shortage.

## The Need for Knowledge and Skills on Drought Management

Drought is part of the weather pattern; it has occurred in the past and will continue to happen in future. So, the organizations concerned with drought management of the drought prone areas need to seek comprehensive and complete data to forecast the likelihood of drought. Drought directly affects water, land and geographical conditions and socio-economy of the locality. The difference between drought and other natural disasters is that its duration is longer than that of others. So, the department concerned need to work in cooperation and coordination to mitigate the drought impact. The measures will include ensuring proper network to be able to gather measurements on meteorology and hydrology and facts that are vital for businesses dependent on weather and water resources; proper exchange of data among the departments to prevent the droughts, mitigate their impact and to respond to them; conducting training of personnel to familiarize them with the data so as to make better use of them in making decisions; making arrangements for farmers and other organizations in order that weather forecasts are useful, clear and simple to understand while minimizing constraints; use of standard rainfall indexes to reliably calculate the beginning and end of droughts; sharing and properly using facts about the drought and weather pattern and working together to be able to have better knowledge about the intensity and the vastness of the areas affected; compiling facts and seeking methods to evaluate the drought impact in order to be able to respond to the ill effect; working harder to see that seasonal weather forecasts reach the local residents and organizations on time; and seeking ways and means to obtain important local and global data on droughts useful to the NGOs and international NGOs. The needs for drought management in Myanmar are as follow:

1. To set up the Forecasting and Warning system for drought management
2. To develop the forecasting techniques and capacity building for drought management
3. To set up a Task force including authorities and experts of Administration, Relief, water resources, Agriculture, Forestry, Meteorological Agency, NGOs, INGOs.
4. To conduct Drought risk assessment
5. To develop a decision support Drought Management
6. To promote Education and public awareness for drought mitigation.
7. To encourage community level plans of Drought Mitigation.
8. To cooperate, coordinate and collaborate the concerned departments and organizations for drought management
9. To develop the concerned department's activities for drought mitigation
10. To develop the drought policy and strategies for drought management in Myanmar

## Conclusions

Drought is a natural hazard that involves many factors, including meteorological and climatological parameters, having complex inter-relationships. Drought definitions vary from region to region and may depend upon the dominating perception, and the task for which it is defined. Identifying patterns of drought and finding its associations with various indices derived from conventional method and remote sensing techniques are becoming important for monitoring of this natural hazard. Dealing with NDVI and rainfall dataset for a time-series of more than 20 year makes the study not only complicated but also difficult to analyze. This study addresses the need of analyzing and studying the pattern of drought using temporal rainfall datasets and their derived indices such as SPI for time-series dataset and vegetation based index (NDVI).

## Acknowledgements

I would like to express my deepest gratitude to my department head Dr. Zin Mar Lwin Remote Sensing Department and suggestions and valuable comments during my studies. I also want to thank Dr. Kyaw Zaya IItun and staffs for the support and help during my studies.

# References

1.  http://www.nationsonline.org/oneworld/myanmar.htm

2.  Tin Yi. Deputy. Director (DMH), Wai Myo Hla. Director (DZGD),Aung Kyaw Htun (DZGD): "Drought Conditions and Management Strategies in Myanmar".

3.  http://reliefweb.int/disaster/dr-2015-000180-vnm.

## Chapter 10

# Drought Management and Desertification Control in Africa: Issues and Mitigation Strategies with Possible Short and Long Term Solutions

*Oseni Adedayo Olusegun*

*Scientific Officer I,*
*Department of Chemical Technology,*
*Federal Ministry of Science and Technology Headquarters, Abuja, Nigeria*
*E-mail: oseniadedayo1@gmail.com*

## Abstract

Drought and desertification are at the core of serious challenges and threats facing sustainable development in Africa. These problems have far reaching adverse impacts on human health, food security, economic activity, physical infrastructure, natural resources, and environment, national and global security.

Although drought has several definitions, the central element in this definition is water deficit. In general, drought is defined as an extended period – a season, a year, or several years – of deficient rainfall relative to the statistical multi-year average for a region. This deficiency results in a water shortage for some activity, group, or environmental sector. Desertification on the other hand is defined as a process of land degradation in arid, semi-arid and dry sub-humid areas, resulting from various factors, including climatic variations and human activities. Drought is a creeping phenomenon with slow onset and which makes it difficult to define when it begins and when it ends.

The aim and objective of this study is to critically look into issues and mitigation strategies with solutions for drought and desertification. The aim is to develop an easy-to-use resource which can be used in decision making for drought risk management.

Drought is a protracted period of deficient precipitation which causes extensive damage to crops, resulting in loss of yield. Operational definitions help identify the drought's beginning, end, and degree of severity. To determine the beginning of drought, operational definitions specify the degree of departure from the precipitation average over some time period. This is usually accomplished by comparing the current situation with the historical average. The threshold identified as the beginning of a drought (*e.g.*, 75 per cent of average precipitation over a specified time period) is usually established somewhat arbitrarily. An operational definition for agriculture may compare daily precipitation to evapotranspiration to determine the rate of soil-moisture depletion, and express these relationships in terms of drought effects on plant behavior. Operational definitions are used to analyze drought frequency, severity, and duration for a given historical period. Such definitions, however, require weather data on hourly, daily, monthly, or other time scales and, possibly, impact data (*e.g.*, crop yield).

### Strategies that Can be Used at Mitigating Drought and Desertification

Encourage ecosystem-based adaption in the drylands and beyond, rehabilitate degraded land to increase climate change resilience, build technical and institutional capacity for sustainable land management and create national and sub-national policies for drought mitigation and prevention.

### Possible Short and Long-term Solutions

Short term solutions for drought and desertification are setting up of refugee camps and assisting with medical aid and also giving food and clothing aid.

Long term solutions for drought and desertification include fencing off areas to prevent animals grazing there, introducing drought resistant crops *e.g.* millet, planting trees and bushes to provide cover for soil and stop it being blown or washed away and rotating crops and water them using irrigation techniques.

*Keywords: Drought, Drought management, Desertification, Environment, Land, Economy.*

## Introduction

Drought and desertification are a complex natural hazard and a twin global environmental problem that occurs in every part of the world and adversely affects the lives of millions of people each year, causing significant damage to economies, the environment, and property (Wilhite, 2014). Africa is faced with rapid desert encroachment affecting most African countries from moderate to severe rate. Drought and desertification impact directly or indirectly on all aspects of human life and the environment including the ecological, health, geo-chemical, hydrological and socio-economic facets. Despite several efforts by the government to end desertification, the problem still persist due to the gap between the formation of policy and strategies of combating drought and desertification. Drought and desertification can be remedied through integrated approaches such as awareness programmes, protection of marginal lands, tree planting, sustainable agricultural practices and use of alternative energy (Olagunju *et al.*, 2015).

Desertification is a land degradation problem of major importance in the arid, semi-arid and dry sub-humid regions of the world. It is one of the most serious resource management problems facing the world today. The United Nations

Convention to Combat Desertification (UNCCD) defined desertification as: "Land degradation in arid, semi-arid and dry sub-humid areas resulting from various factors including climate variations and human activities." Land in this context includes soil, land surface vegetation and local water resources and degradation means reduction of the current and/or the potential productive capacity of the land.

Drought and desertification are global environmental problems affecting developed and developing countries in many regions of the world and they are accompanied by the reduction in the natural potential of the land, the depletion of surface and groundwater and negative repercussions on the living conditions and the economic development of the people are affected by it (Abahussain *et al.*, 2002). Drought and desertification processes integrate climatic elements with human activities in transforming productive land, into an ecological impoverished area generally refers to as desert. Drought and desertification cause degradation of once a fertile land through long term changes in the soil, climate and biota, which results in desert-like conditions.

Drought is one of the main causes of desertification. The lack of general acceptance of a precise and objective definition of drought has been one of the principal obstacles to the investigation of drought. It is therefore important to be aware that different definitions might lead to different conclusions regarding the drought phenomenon. For instance, if the definition is based on the level of rainfall, it is possible that rainfall statistics summarized over a calendar year indicate no drought, whereas the moisture supply in the growing season does. With regards to food security, drought could be defined as naturally occurring phenomenon (usually aggravated by human activities) that exist over a particular period in a particular area such that precipitation is significantly below normal recorded levels, causing deterioration of land productive systems and invariably low agricultural outputs.

## Literature Review

### Global Status on Drought and Desertification

According to the United Nations statistics deterioration of soil and plant cover, it has adversely affected 70 per cent of the world's dry lands or 3.6 billion hectares or a quarter of the total land surface. The impact is staggering more than 250 million of the earth's inhabitants who are directly affected by desertification, 135 million are in danger of being driven from their lands. The livelihoods of one billion people, nearly one fifth of the world's population are at risk. More than 110 million countries have land at risk of desertification. The world wide price tag for desertification exceeds 42 billion US dollars a year (UNEPA, 1992). Such magnitude of the process has put the problem of desertification into focus and desertification was recognized as a global threat in the United Nations Conference on Environment and Development.

### Regional Status on Drought and Desertification

Desertification has its greatest impact in Africa. According to UNSO, two thirds of the continent or 70 per cent is desert or degraded to some degree. Desertification affects half the inhabitants of the continent or about 326 million inhabitants (UNSO,

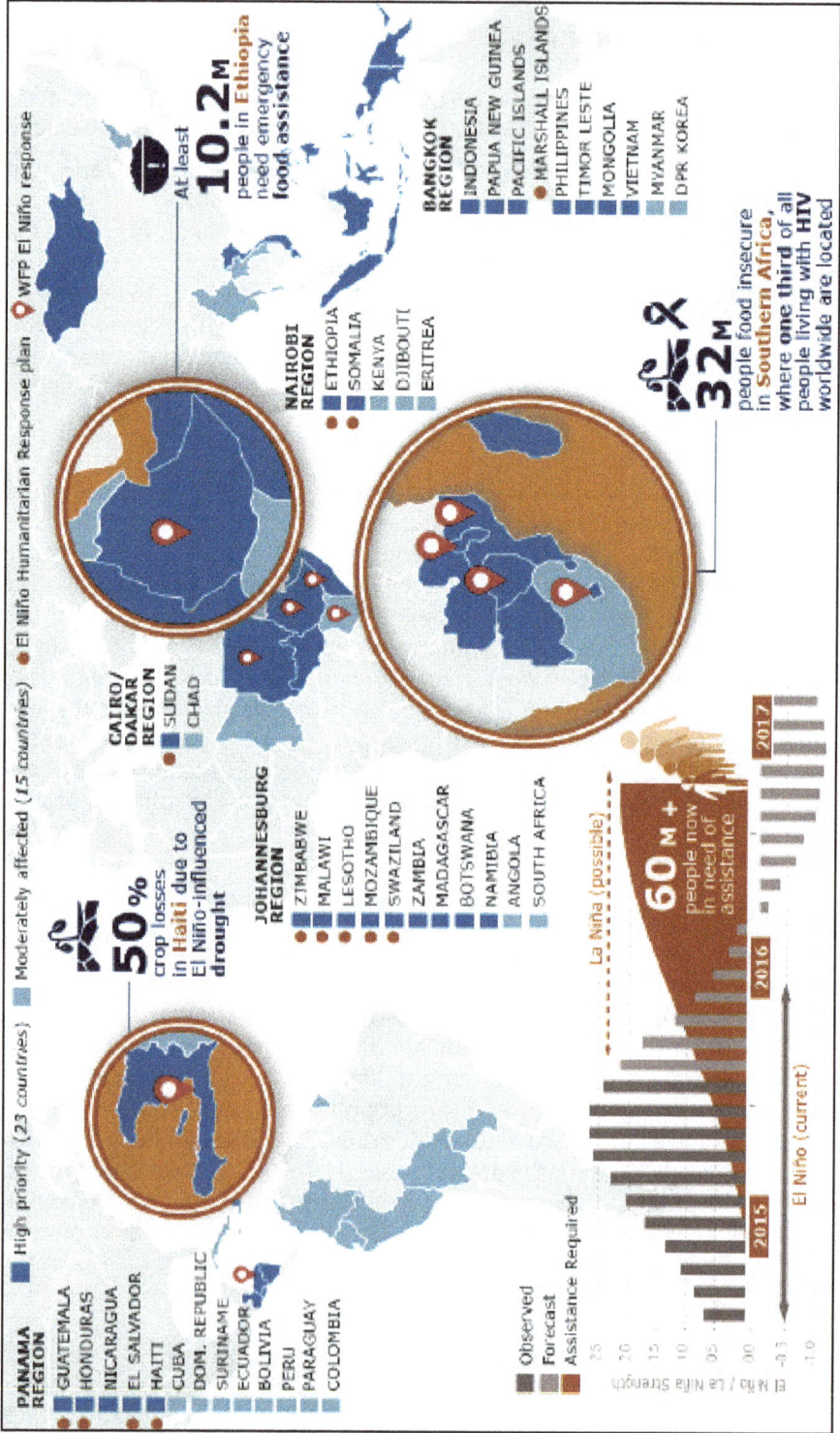

**Figure 10.1: Recent Drought in Africa (as of April 2016) El Nino-related Food Security Impact in the World (*Source*: WFP, 2016b).**

1997). The region is afflicted by frequent and severe droughts. Many African countries are land-locked, have widespread poverty, need external assistance, and depend heavily on natural resources for subsistence. In these countries, combating desertification and promoting development are virtually the same due to the social and economic importance of natural resources and agriculture. When people live in poverty, they have little choice but to over-exploit the land. (UNEPA, 1992)

## Status and Trends of Drought and Desertification in Africa

Two thirds of Africa is classified as deserts or dry lands (UNESCO, 2007). These are concentrated in the Sahelian region, the Horn of Africa and the Kalahari in the south. Africa is especially susceptible to land degradation and bears the greatest impact of drought and desertification. It is estimated that two-thirds of African land is already degraded to some degree and land degradation affects at least 485 million people or sixty-five per cent of the entire African population. Desertification especially around the Sahara has been pointed out as one of the potent symbols in Africa of the global environment crisis. Climate change is set to increase in the area susceptible to drought, land degradation and desertification in the region. Under

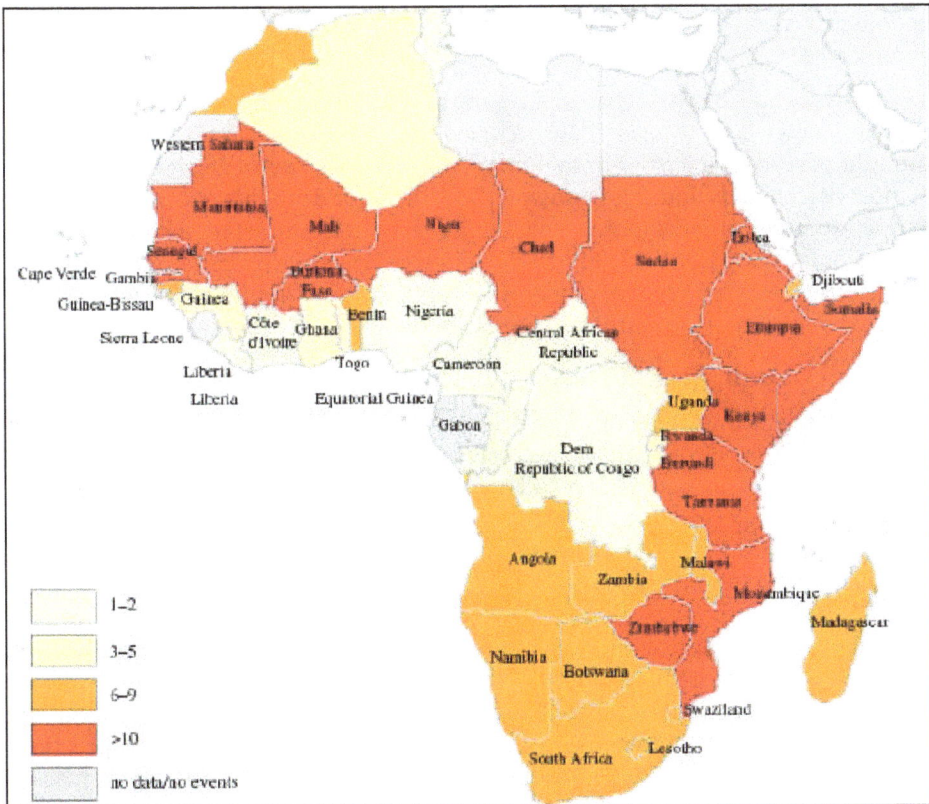

**Figure 10.2: Drought Events per Country from 1970 to 2004 within Sub-Saharan Africa (*Source*: Noojin, Leah 2006. Factors that influences famine in Sub-Saharan African Countries).**

a range of climate scenarios, it is projected that there will be an increase of 5-8 per cent of arid and semi-arid lands in Africa. It is also estimated that 35 per cent of the land area (about 83,489 km2 or 49 out of the 138 districts) of Ghana is prone to desertification (UNESCO, 2007), with the Upper East Region and the eastern part of the Northern Region are facing the greatest hazards. Indeed a recent assessment indicates that the land area prone to desertification in the country has almost doubled during recent times. Desertification is said to be creeping at an estimated 20,000 hectares per year, with the attendant destruction of farmlands and livelihoods in the country. Seventy per cent of Ethiopia is reported to be prone to desertification, while in Kenya, around 80 per cent of the land surface is threatened by desertification. Estimates of the extent of land degradation within Swaziland suggest that between 49 and 78 per cent of the land is at risk, depending on the assessment methodology used (Government of Swaziland, 2000). Nigeria is reported to be losing 1,355 square miles (1mile =1.6km) of rangeland and cropland to desertification each year and this affects each of the 10 northern states of Nigeria. It is estimated that more than 30 per cent of the land area of Burundi, Rwanda, Burkina Faso, Lesotho and South Africa is severely or very severely degraded. These rates and extent of land degradation/ desertification undermine and pose serious threats to livelihoods of millions of people struggling to edge out of poverty. With regard to drought, the continent has witnessed a high frequency of occurrence and severity of drought as shown below.

Drought is one of the most important climate-related disasters in Africa. Climate change is set to exacerbate occurrence of climate related disasters including drought. A study from Bristol University projects that areas of western Africa were at most risk from dwindling freshwater supplies and droughts as a result of rising temperatures. Current climate scenarios predict that the driest regions of the world will become even drier, signalling a risk of persistence of drought in many parts of Africa (arid, semi-arid and dry sub humid areas) which will therefore bear greater and sustained negative impacts. (UNESCO, 2007)

## Causes of Drought and Desertification

The underlying cause of most droughts can be related to changing weather patterns manifested through the excessive build-up of heat on the earth's surface, meteorological changes which result in a reduction of rainfall, and reduced cloud cover, all of which results in greater evaporation rates. The resultant effects of drought are exacerbated by human activities such as deforestation, overgrazing and poor cropping methods, which reduce water retention of the soil, and improper soil conservation techniques, which lead to soil degradation. Desertification is caused by multiple direct and indirect factors. It occurs because dry lands ecosystems are extremely vulnerable to over-exploitation and inappropriate land use that result in underdevelopment of economies and in entranced poverty among the affected population. Whereas over cultivation, inappropriate agricultural practices, overgrazing and deforestation have been previously identified as the major causes of land degradation and desertification, it is in fact a result of much deeper underlying forces of socio-economic nature, such as poverty and total dependency on natural resources for survival by the poor. (UNESCO, 2007)

The causes of drought and desertification are numerous and complex, but like many other issues of environmental degradation, they are basically the resultant interactions of climatic influence and human activities in the environment. The causes include:

## Climatic Variability

Climatic variability is a major driver of many environmental degradation phenomena. Alteration of climatic conditions leads to naturally occurring phenomena of drought and desertification. There has been increasing level of greenhouse gases causing global warming which in turn increase the variability of climate conditions. This alteration in the climatic conditions has manifested as follows:

1. A decrease in the amount of rainfall in dry lands making arid and semi-arid lands more vulnerable to desertification.
2. High temperatures, combined with low rainfall which would lead to the drying up of water resources - drought.
3. Poor growth of vegetation leading to the formation of a desert-like condition.

## Anthropogenic Activities

The anthropogenic factors have been the major cause of desertification just like many ecological degradation problems. Human contribute to desertification through poor land use and the ever increasing pressure put upon the limited available resources by the expanding population. Basically, human causes of desertification can be viewed to result from; exploitation of resources from ¯non-ideal lands, over exploitation of land resources, unsustainable acts when exploiting, and none replacement of exploited resources or not allowing sufficient time for natural regeneration of exploited resources. The following human activities can cause desertification:

### i. Deforestation

Deforestation is the conversion of forested areas to non-forested land (Olagunju, 2015a). It is the large scale removal of forests resulting to non-forest to meet various human needs. Logging, expansion of agricultural croplands, urbanization, fuel wood collection, mining and resources extraction, fire-hunting and slash and burn practices have been identified as the key drivers of deforestation. Deforestation of dry lands destroys the trees and vegetation that bind the soil, and because of the prevailing climatic conditions in dry lands, the possibility of regeneration of denuded vegetation is low and hence, the land becomes desertified.

### ii. Extensive Cultivation

Expansion of agricultural land to meet up with the food requirements of the increasing population has led to the degradation of land in Africa (Temidayo, 2015). New lands are cleared of trees and other vegetations to establish agricultural croplands in the dry land, many of such lands are unable of recuperation, and hence desertification sets in. In Nigeria for example, overgrazing and over-cultivation have been reported to be responsible for the conversion of 351,000 hectares of land into desert each year.

### iii. Overgrazing

Overgrazing is most common in the areas whose socioeconomic viability depend mostly on extrinsic system of animal husbandry. The dry lands of Nigeria is said to support much of the country's livestock economy, hosting about 90 per cent of the cattle population, about two-thirds of the goats and sheep and almost all donkeys, camels and horses. In the Sudan and the Sahel zones, which carry most of the livestock population, nomadic herdsmen graze their livestock throughout the area and are constantly in search of suitable pastures. Additional pressure is also on these natural rangelands by livestock from neighbouring countries, notably Cameroon, Chad and Niger. Overgrazing removes the vegetation cover that protects soil from erosion (UNCCD, 2011) and degrades natural vegetation that leads to desertification and decrease in the quality of rangelands (Sheikh *et al.*, 2006).

### iv. Cultivation of Marginal Land

Cultivation of marginal areas is one of the causes of desertification. Marginal lands are areas that are unable to support permanent or intensive agriculture which could be easily degraded following cultivation. During the periods of high rainfall, people tend to extend farming activities into the marginal areas. When these periods of high precipitation is succeeded by abrupt dry periods, the exposed land with very little vegetal cover is prone to wind erosion and desertification may set in which could be irreversible except through carefully planned rehabilitation programme.

### v. Bush Burning

Slash and burn practice in agriculture and fire-hunting is a major cause of desertification in Africa (Ebenezer, 2012). Owing to the low relative humidity in the area coupled with very dry harmattan wind, there is always a high incidence of bush fires every dry season. When this occurs too frequently, the vegetation may not regenerate; the soil is exposed to erosion and become degraded.

### vi. Fuel Wood Extraction

Due to socio-economic status of the people inhabiting Nigeria dryland, felling of tree for fuel wood will continue increasing if alternative sources of energy in the sudano-sahelian zone are not provided. The demand for fuel wood causes the removal of trees, shrubs, herbaceous plants and grass cover from the fragile land, accelerating the degradation of the soil to desert-like conditions (FAO, 2006). In Nigeria for example, more than 70 per cent of the nation's population depends on fuel wood. Katsina alone, a northern state, has its over 90 per cent energy from fuel wood (Mohammed *et al.*, 2013). In Kano City, 75,000 tonnes of fuel wood are brought in by lorry and donkey within a radius of 20 km, which leads to denuding of the woodland.

### *vii. Faulty Irrigation Management*

Irrigation system is a common practice in northern Nigeria. Many farmers lack adequate skills in proper designing and management of irrigation system which has resulted into desert-like condition of many irrigated farmlands as a result of water logging and salinization. This scenario is already a reality on a number of irrigation projects in Nigeria today, such as the Bakolori Irrigation, South Chad Irrigation and the Hadejia – Jamaare Irrigation Projects.

### *viii. Urbanization*

Nneji (2013) has attributed rapid economic growth and urbanization as causal factors of desertification. The problem is more severe and complicated in developing world. Clearing of lands to accommodate the increasing population and accommodate the necessary infrastructure is commonly done without adequate environmental consideration; this has led to the removal of vegetation cover in the area and as such, making the area desertified.

### *ix. Natural Disasters*

There are some cases where the land gets damaged because of natural disasters, including drought. In those cases, there isn't a lot that people can do except work to try and help rehabilitate the land after it has already been damaged by nature.

## Major Impacts of Drought

Drought produces a complex web of impacts that spans many sectors of the economy and reaches well beyond the area experiencing physical drought. This complexity exists because water is integral to society's ability to produce goods and provide services.

Impacts are commonly referred to as direct and indirect. Direct impacts include reduced crop, rangeland, and forest productivity, increased fire hazard, reduced water levels, increased livestock and wildlife mortality rates, and damage to wildlife and fish habitat. The consequences of these direct impacts illustrate indirect impacts. For example, a reduction in crop, rangeland, and forest productivity may result in reduced income for farmers and agribusiness, increased prices for food and timber, unemployment, reduced tax revenues because of reduced expenditures, foreclosures on bank loans to farmers and businesses, migration, and disaster relief programs (Salami, 2007).

The types of drought impacts are described below.

## Economic Impacts

Many economic impacts occur in agriculture and related sectors, because of the reliance of these sectors on surface and groundwater supplies. In addition to losses in yields in both crop and livestock production, drought is associated with insect infestations, plant disease, and wind erosion. The incidence of forest and range fires increases substantially during extended periods of droughts, which in turn places both human and wildlife populations at higher levels of risk.

## Environmental Impacts

Environmental losses are the result of damages to plant and animal species, wildlife habitat, air and water quality, forest and range fires, degradation of landscape quality, loss of biodiversity, and soil erosion. Some of these effects are short-term, conditions returning to normal following the end of the drought. Other environmental effects last for some time and may even become permanent. Wildlife habitat, for example, may be degraded through the loss of wetlands, lakes, and vegetation. However, many species eventually recover from this temporary aberration. The degradation of landscape quality, including increased soil erosion, may lead to a more permanent loss of biological productivity.

## Social Impacts

Social impacts involve public safety, health, conflicts between water users, reduced quality of life, and inequities in the distribution of impacts and disaster relief. Many of the impacts identified as economic and environmental have social components as well. Population migration is a significant problem in many countries, often stimulated by a greater supply of food and water elsewhere. Migration is usually to urban areas within the stressed area, or to regions outside the drought area. Migration may even be to adjacent countries.

## Consequences of Drought and Desertification

Desertification is threatening all the potentially cultivable land in many African countries. It leads to many problems. It affects the physical, social and economic life of the affected communities. The loss of soil fertility and vegetation cover and decline of groundwater, which characterized desertification, lead to loss of biodiversity and productivity and contribute to the social, economic and political tensions. The lower yield of crops and the limited grazing area for animals can lead to famines, poverty, death of livestock and conflicts. This may force people to move away from their local areas and consequently lose their connection to their lands and affect their cultural traditions. The displacement of affected communities may also put pressure on other fragile environments and cause conflict and further relocations.

The environmental resources in and around the cities and camps where the displaced people settle are facing severe pressure and the difficult living conditions beside the loss of cultural identity further undermine social stability. (Mortimore, 1989).

Desertification is both a cause and consequence of poverty which at the end of the day forced the affected communities to seek their livelihoods from fragile resources and cause endemic degradation. This is because of extended droughts, human mismanagement of cultivated, and forests and rangelands. As a result of the high population growth rates of human and animals, accompanied with land use conflicts the situation is worsen.

Economically, desertification depletes the natural resources; especially in Sahelian countries where the gross domestic product (GDP) is always low. Though the economic and social costs outside the affected areas, including the influx of

| Various aspects of the economy | Economic structure | | | |
|---|---|---|---|---|
| | Simple | Intermediate | Complex | Dual/extractive |
| Per capita income | Low | Low/low-middle | High-middle/high | Low/middle/high |
| Main sector | Agriculture | Agriculture, manufacturing | Manufacturing, service | Manufacturing, service, agriculture |
| Importance and nature of agriculture sector | Mainly rainfed, accounting for > 20 percent of GDP and > 70 percent of employment | Rainfed/irrigated, accounting for > 20 percent of GDP and 50 percent of employment | Mainly irrigated, accounting for < 10 percent of GDP and < 20 percent of employment | Rainfed/irrigated, accounting for 10-20 percent of GDP and 20-50 percent of employment |
| Intersectoral linkages | Weak | Intensive | Diffused | Weak |
| Engine of growth | Agriculture | Agriculture/non-agriculture | Non-agriculture | Extractive sector |
| Infrastructure | Limited | Extensive | Extensive | Limited/extensive |
| Spatial impact of drought | Largely rural, area directly affected | National, rural and urban population | Largely rural, area directly affected | Rural |
| Economic recovery following drought | Relatively fast | Agriculture - relatively fast; Manufacturing - more slowly | Agriculture - relatively fast; Manufacturing - no real impact | Agriculture; Limited knock-on affects |

**Figure 10.3: Impacts of Drought on different Economic Structures (*Source:* Africa Drought Conference, 2012).**

displaced, refugees and losses to national food production are very high and reflect a huge drain on economic resources.

## Remedies to Drought and Desertification

Solution to the problem of desertification must target all aspects that relate to the problem. Though some desert conditions are irreversible even if all anthropogenic causes are stopped, but some are reversible. (Mumoki, 2006). Some of the remedies to desertification include:

### Awareness

Raising awareness of desertification at local, national and global level is key to remedying drought and desertification. It is probably the cheapest means in combating desertification because it serves as a preventive measure. Awareness will provide people with the understanding of the causes and consequences of the phenomena so as to stop all possible causes and encourage actions that would remedy some of the consequences and prevent further degradation of soil.

### Protection of Marginal Lands

Due to the incapability of marginal lands to support permanent or intensive agriculture, there is need for proper evaluation of such lands with government policy and enforcement aimed at protecting them from any activities that is capable of denuding its vegetation cover.

### Planting and Protection of Indigenous Tree and Shrub Species

Increasing the area of conspicuous vegetation into desertifying lands is vital in managing desertification. This could be done through intensive and technologically supportive reclamation, by planting and establishing indigenous trees and vegetation known to the area. Planting of trees coupled with avoided felling should be embraced in arid and semi-arid zones until if possible a forest zone is attained (Mumoki, 2006). Planting of tress helps in:

1. Soil stability
2. Protection of soil from erosion
3. Retention of soil moisture and nutrients
4. Carbon sequestration

### Sustainable Agricultural Practices

Agroforestry is a form of farming system that plays an extremely important role in the land management of semi-arid and arid zones. Agroforestry is a land use management system in which trees or shrubs are grown around or among crops or pastureland. It combines agricultural and forestry techniques to create more diverse, productive, profitable, healthy, and sustainable land-use system. Grazing systems should be improved from denuding the natural rangelands whose consumption will lead to aridity condition hence establishment of new pastures for grazing by livestock should be ensured. All water to be used for irrigation should be examined to be devoid of level of salt that could result in salt accumulation, as well as ensuring a good drainage system (Sultana, 2008).

## Use of Alternative Source of Energy

Feeling of the few trees and shrubs in desert-prone areas for fuel wood can be reduced through the development of sustainable alternative energy sources such as biofuel. This will not only conserve forest resources but will reduce environmental pollution.

## Effects of Drought and Desertification

★ **Farming becomes next to impossible:** If an area becomes a desert, then it's almost impossible to grow substantial crops there without special technologies. This can cost a lot of money to try and do, so many farmers will have to sell their land and leave the desert areas.

★ **Hunger:** Without farms in these areas, the food that those farms produce will become much scarcer, and the people who live in those local areas will be a lot more likely to try and deal with hunger problems. Animals will also go hungry, which will cause even more of a food shortage.

★ **Flooding:** Without the plant life in an area, flooding is a lot more eminent. Not all deserts are dry; those that are wet could experience a lot of flooding because there is nothing to stop the water from gathering and going all over the place.

★ **Poor Water Quality:** If an area becomes a desert, the water quality is going to become a lot worse than it would have been otherwise. This is because the plant life plays a significant role in keeping the water clean and clear; without its presence, it becomes a lot more difficult.

★ **Overpopulation:** When areas start to become desert, animals and people will go to other areas where they can actually thrive. This causes crowding and overpopulation, which will, in the long run, end up continuing the cycle of desertification that started this whole thing anyway.

★ **Poverty:** Drought and desertification can lead to poverty if it is not kept in check. Without food and water, it becomes harder for people to thrive, and they take a lot of time to try and get the things that they need.

## Results and Discussion

Drought definitions are of two types: (1) conceptual, and (2) operational. Conceptual definitions help understand the meaning of drought and its effects. For example, drought is a protracted period of deficient precipitation which causes extensive damage to crops, resulting in loss of yield. Operational definitions help identify the drought's beginning, end, and degree of severity. To determine the beginning of drought, operational definitions specify the degree of departure from the precipitation average over some time period. This is usually accomplished by comparing the current situation with the historical average. The threshold identified as the beginning of a drought (*e.g.*, 75 per cent of average precipitation over a specified time period) is usually established somewhat arbitrarily.

An operational definition for agriculture may compare daily precipitation to evapotranspiration to determine the rate of soil-moisture depletion, and express these relationships in terms of drought effects on plant behavior. Operational definitions are used to analyze drought frequency, severity, and duration for a given historical period. Such definitions, however, require weather data on hourly, daily, monthly, or other time scales and, possibly, impact data (*e.g.*, crop yield). Climatology of drought for a given region provides a greater understanding of its characteristics and the probability of recurrence at various levels of severity. The various types of droughts are listed below.

## Meteorological Drought

Meteorological drought is defined on the basis of the degree of dryness, in comparison to a normal or average amount, and the duration of the dry period. Definitions of meteorological drought must be region-specific, since the atmospheric conditions that result in deficiencies of precipitation are highly region-specific.

The variety of meteorological definitions in different countries illustrates why it is not possible to apply a definition of drought developed in one part of the world to another. For instance, the following definitions of drought have been reported:

* ★ United States (1942): Less than 2.5 mm of rainfall in 48 hours.
* ★ Great Britain (1936): Fifteen consecutive days with daily precipitation less than 0.25 mm.
* ★ Libya (1964): When annual rainfall is less than 180 mm.
* ★ Bali (1964): A period of six days without rain.

Data sets required to assess meteorological drought are daily rainfall information, temperature, humidity, wind velocity and pressure, and evaporation.

## Agricultural Drought

Agricultural drought links various characteristics of meteorological drought to agricultural impacts, focusing on precipitation shortages, differences between actual and potential evapotranspiration, soil-water deficits, reduced groundwater or reservoir levels, and so on. Plant water demand depends on prevailing weather conditions, biological characteristics of the specific plant, its stage of growth, and the physical and biological properties of the soil. A good definition of agricultural drought should account for the susceptibility of crops during different stages of crop development. Deficient topsoil moisture at planting may hinder germination, leading to low plant populations per hectare and a reduction of yield.

## Hydrological Drought

Hydrological drought refers to a persistently low discharge and/or volume of water in streams and reservoirs, lasting months or years. Hydrological drought is a natural phenomenon, but it may be exacerbated by human activities. Hydrological droughts are usually related to meteorological droughts, and their recurrence interval varies accordingly. Changes in land use and land degradation can affect the magnitude and frequency of hydrological droughts.

## Socioeconomic Drought

Socioeconomic definitions of drought associate the supply and demand of some economic good with elements of meteorological, hydrological, and agricultural drought. It differs from the other types of drought in that its occurrence depends on the processes of supply and demand. The supply of many economic goods, such as water, forage, food grains, fish, and hydroelectric power, depends on the weather. Due to the natural variability of climate, water supply is ample in some years, but insufficient to meet human and environmental needs in other years.

Socioeconomic drought occurs when the demand for an economic good exceeds the supply as a result of a weather-related shortfall in water supply. The drought may result in significantly reduced hydroelectric power production because power plants were dependent on streamflow rather than storage for power generation. Reducing hydroelectric power production may require the government to convert to more expensive petroleum alternatives, and to commit to stringent energy conservation measures to meet its power needs.

The demand for economic goods is increasing as a result of population growth and economic development. The supply may also increase because of improved production efficiency, technology, or the construction of reservoirs. When both supply and demand increase, the critical factor is their relative rate of change. Socioeconomic drought is promoted when the demand for water for economic activities far exceeds the supply.

## Solutions for Drought and Desertification

* **Policy Changes Related to How People can Farm:** In countries where policy change will actually be enforced on those in the country, policy change related to how often people can farm and how much they can farm on certain areas could be put into place to help reduce the problems that are often associated with farming and desertification.

* **Policy Changes to Other Types of Land Use:** If people are using land to get natural resources or they are developing it for people to live on, then the policies that govern them should be ones that will help the land to thrive instead of allowing them to harm the land further. The policy changes could be sweeping or they could be depending on the type of land use at hand.

* **Education:** In developing countries, education is an incredibly important tool that needs to be utilized in order to help people to understand the best way to use the land that they are farming on. By educating them on sustainable practices, more land will be saved from becoming desert.

* **Technology Advances:** In some cases, it's difficult to try and prevent desertification from happening. In those cases, there needs to be research

and advancements in technology that push the limits of what we currently know. Advancements could help us find more ways to prevent the issue from becoming epidemic.

★ **Putting Together Rehabilitation Efforts:** There are some ways that we can go back and rehabilitate the land that we've already pushed into desertification; it just takes some investment of time and money. By putting these together, we can prevent the issue from becoming even more widespread in the areas that have already been affected.

★ **Sustainable practices to prevent desertification from happening:** There are plenty of sustainable practices that can be applied to those acts that may be causing desertification.

## Mitigation Strategies

Drought is a natural hazard, it has a slow onset, and it evolves over months or even years. It may affect a large region and causes little structural damage. The impacts of drought can be reduced through preparedness and mitigation.

The components of a drought preparedness and mitigation plan are the following:

★ Prediction

★ Monitoring

★ Impact assessment

★ Response.

**Figure 10.4: Disaster, Risk and Crisis Management Cycle (*Source*: Wilhite, 1999).**

Prediction can benefit from climate studies which use coupled ocean/ atmosphere models, survey of snow packs, anomalous circulation patterns in the ocean and atmosphere, soil moisture, assimilation of remotely sensed data into numerical prediction models, and knowledge of stored water available for domestic, stock, and irrigation uses.

Monitoring exists in countries which use ground-based information such as rainfall, weather, crop conditions and water availability. Satellite observations complement data collected by ground systems. Satellites are necessary for the provision of synoptic, wide-area coverage.

Impact assessment is carried out on the basis of land-use type, persistence of stressed conditions, demographics and existing infrastructure, intensity and areal extent, and its effect on agricultural yield, public health, water quantity and quality, and building subsidence.

Response includes improved drought monitoring, better water and crop management, augmentation of water supplies with groundwater, increased public awareness and education, intensified watershed and local planning, reduction in water demand, and water conservation.

## Conclusions

Drought and desertification continue to threaten the livelihoods of millions of people in Africa, increasingly making them unable to edge out of poverty. This trend is set to worsen with the set of climate change, to which many countries in the region are most vulnerable. As such, desertification and drought are at the heart of development challenges in Africa and needs urgent attention in policies and actions at national regional and global levels. Combating desertification in the continent has tremendous benefits in enhancing the continent's progress towards meeting MDGs particularly in terms of poverty reduction, attaining food security, combating diseases and ensuring environmental sustainability.

African countries continue to face a number of challenges and constraints that constitute major impediments to and hamper progress in addressing drought and desertification and attaining sustainable development. Notable are the high levels of poverty, weak institutional capacities, and challenges in resource mobilisation, weak information base, and inadequate access to affordable appropriate technology. In order to succeed and achieve significant progress in combating desertification and mitigating impacts of drought, there is need for enhanced political will and commitment at all levels to address these problems as an integral priority.

## Recommendations

   ✶ Creation of National programmes of priorities such as the establishment of the national coordination body, the establishment of the national desertification fund.

   ✶ Raising awareness at community level to succeed their participation and involvement in combating desertification.

★ Each state should develop legislation and/or enforce laws that prohibit over exploitation of natural resources and encourage the protection and conservation of terrestrial natural resources.

★ Raising awareness and capacity building of institutions and organizations that deal with the natural resources.

★ Traditional knowledge and research should be oriented in each state to address the problem of desertification.

★ Monitoring the status of desertification needs networking with different institutions and organizations and capacity building and sharing of information, so it is recommended that networking between the states and centre and relative organizations should be built urgently.

## Acknowledgements

It is my sincere wish and gratitude to express my profound appreciation to the Almighty God for guidance and protection in making my dream come true.

My immense appreciations go to the Director (Planning, Research and Policy Analysis – Focal Point), Mr. Ibrahim Suleiman, Deputy Director (Chemical Technology Department) Mr. Akinyemi Oyefeso and my other colleagues who I am unable to mention here for their contributions.

I also wish to acknowledge the efforts of the organizers: Centre for Science and Technology of the Non-Aligned and Other Developing Countries (NAM S&T Centre) and the Ferdowsi University of Mashhad, Iran for considering me to participate in the **International Workshop on Drought Management and Desertification Control. I am very grateful.**

Finally, I must also appreciate and thank my mum for her care, support, love and prayers to see that I took part in this programme. I equally appreciate the prayers of my siblings, friends and loved ones.

## Abbreviations

GDP: Gross Domestic Product

UNESCO: United Nations Educational, Scientific and Cultural Organization

UNCCD: United Nations Convention to Combat Desertification

FAO: Food and Agriculture Organization

SNAP: Sudan National Action Programme

UNSO: United Nations Sudano-Sahelian Office (UNDP)

UNEP: United Nations Environment Programme

UNDP: United Nations Development Programme

MDG: Millennium Development Goals

# References

1.   Abahussain AA, Abdu A, Al-Zubari WK, El-Deen NA, Abdul-Raheem M (2002). Desertification in the Arab Region: analysis of current status and trends. *J. Arid Environ.* 51: 521-545.

2.   Ebenezer (2012). Environmental Science: Effect of Drought in the Environment: 32-34.

3.   FAO (2006). Global Forest Resources Assessment 2005. Progress towards sustainable forest management. FAO Forestry Paper No. 147, p. 320.

4.   Olagunju (2015a). Impacts of Human-induced Deforestation, Forest Degradation and Fragmentation on Food Security. *New York Sci. J.* 8(1): 4-16.

5.   Mortimore M (1989). Adapting to drought: farmers, famines, and desertification in West Africa. Cambridge University Press. pp.12. ISBN 978-0-521-32312-3.

6.   Mohammed D, Akpan AE, Aliyu HS (2013). Role of Community Participation in Combating Desertification in the Arid Zone of Nigeria: An Overview. *J. Environ. Manage. Saf.* 4(3): 49-58.

7.   Mumoki F (2006). The effects of deforestation on our environment today. http://www.Tigweb.org/express/panorama/articles.html?

8.   Nneji LM (2013). A Review of the Effects of Desertification on Food Security. *Rep. Opin.* 5(10): 27-33.

9.   Salami (2007). Environmental Studies: The Major impacts of Drought 19(3): 10-14.

10.  Sheikh BA, Soomro GH (2006). Desertification: Causes, Consequences and Remedies. *Pak. J. Agric. Agric. Eng. Vet. Sci.* 22(1): 44-51.

11.  Sultana S (2008). Excessive Irrigation Promotes Desertification. Google/The New Nation of the June 28, 2008.

12.  Temidayo (2015). Desertification Threatens Economy, *Food Security*. 301-306.

13.  Wilhite, D.A. and Glantz, M.H., (1985). Understanding: the drought phenomenon: the role of definitions. *Water International*, 10(3): 111-120.

14.  Wilhite, D.A., Sivakumar, M.V. and Pulwarty, R., (2014). Managing drought risk in a changing climate: The role of national drought policy. *Weather and Climate Extremes*, 3: 4-13.

15.  UNCCD (2011). Desertification: A visual Synthesis. GRAPHI 4 Press, Bresson France. pp. 1-52.

16.  UNEPa (1992). Status of desertification and implementation of the United Nations Plan of Action to Combat Desertification, Report of the Executive Director, Nairobi, United Nations Environment Programme.

17.  UNSO (1997). Aridity zones and Dryland Populations, Office to Combat Desertification and Drought (UNSO/UNDP), September 1997, Printed and recycled paper, UNDP, One United Nations Plaza, New York, NY 10017.

18. UNESCO (2007). Economic Commission for Africa, Fifth Meeting of the Africa Committee on Sustainable Development (ACSD-5) Regional Implementation Meeting (RIM) for CSD-16 Addis Ababa 22-25 October 2007.

# Chapter 11

# Drought Management and Desertification Control in Northern Nigeria: A Review

*Abubakar Shehu Kollere*

*Raw Materials Research and Development Council,*
*17, Aguiyi Ironsi Street, Maitama, Abuja Nigeria*
*E-mail: kkollere@yahoo.com*

## Abstract

Drought and desertification constitute the two greatest challenges to the socio-economic life of the people of Northern Nigeria. Some parts of the region lie in the Sudano-Sahel and the southern fringes of the Sahara Desert. The paper attempts to highlight the dangers of drought and desertification and its effects on the socio-economic life of the people of Northern Nigeria and government efforts in combating and or mitigating its impacts. Drought and desertification directly or indirectly affect the socio-economic life of the people ranging from diseases such as polio and meningitis to food in-security and conflicts as a result of religious, ethnic and tribal clashes such as the Boko-Haram and the perennial attacks between farmers and cattle herders. Fifteen (15) of the 19 states of the north are affected at various degrees of desert encroachment. Factors responsible for this include climatic variability and human activity which include deforestation, overgrazing, cultivation of marginal soils and urbanization. Some of the environmental effects include increase exposure to drought, reduced water table, loss of livestock, farmer-cattle rearer conflicts, *etc*. The government efforts at combating drought and desertification include creation awareness on the dangers of drought and desertification on electronic and print media, annual tree planting campaigns, the Great Green Wall project and promoting alterative, sustainable and more efficient energy sources. However, more synergistic efforts are needed among critical stakeholders in order to ameliorate the effects of drought and desertification.

*Keywords: Drought, Desertification, Climate change, Control, Northern Nigeria.*

## Introduction

The Northern Nigeria is characterized by arid and semi-arid to desert climatic conditions due to its location at the extreme northern end of the country in the Sahel region bordering the Sahara Desert. Its people are mostly farmers, livestock herders and fishermen, who are found in the shores of the Lake Chad and along the banks of rivers Niger and Benue and their tributaries (Figure 11.1).

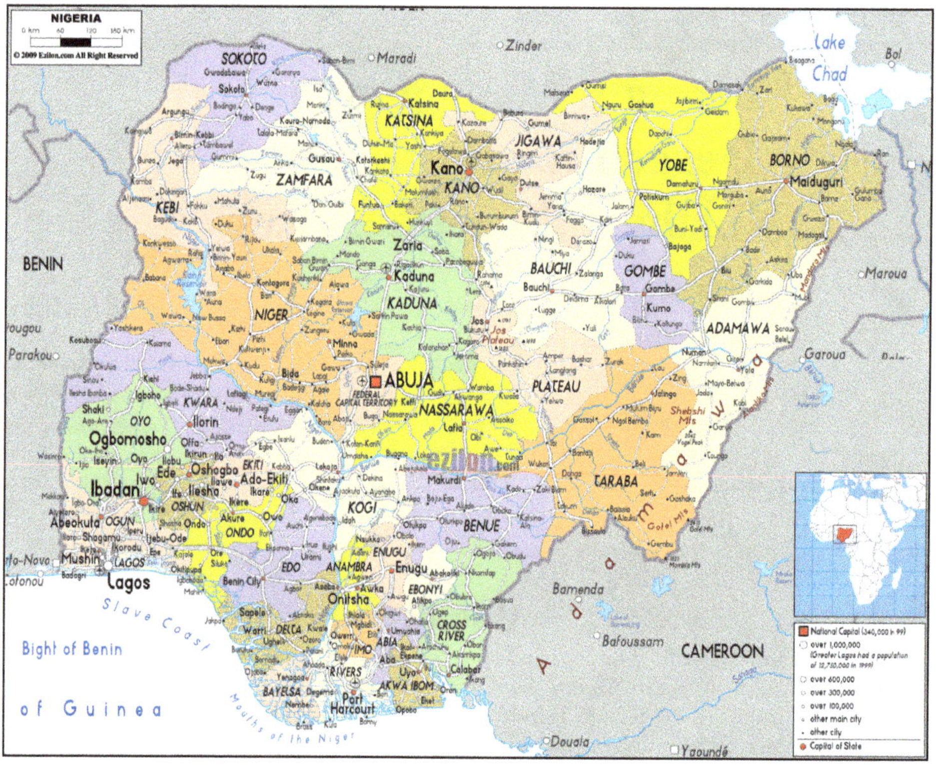

**Figure 11.1: Political Map of Nigeria Showing 36 States and the Federal Capital Territory.**

Drought and desertification has been described as twin global environmental problems. These are a cause of a reduction in the natural potential of the land, surface and groundwater and together have negative consequences on the living conditions and the economic development of the people it is affecting. Desertification has been associated to Northern Nigeria since in the 1920s but its adverse impacts were felt greatly during the 1968-1973 drought in the region that led to famine and loss of human and livestock lives.

The Northern states are also known to be the food basket of Nigeria with crops like, sorghum, millet, melon, maize, rice and sesame seed, and wheat to a lesser extent. The horticultural crops include tomatoes, onions, peppers and variety of other vegetables. The region also produces legumes such as cow pea, groundnuts, soy bean, *etc*. Tree crops found in the region include those that occur in the wild or are cultivated in plantations which include acacias that produce gum Arabic, Shea nuts and mangoes. The Northern states also provide most of the livestock need of Nigeria such as fishery, cattle, sheep and goats and their dairy and associated products of hide and skin and leather which are major foreign exchange earners.

In spite of the water resources availability of Nigeria, the impacts of drought and desertification is felt because agriculture is mostly supported by the seasonal rain that is sometimes less than 500mm in the extreme northern states and over 2500mm in the Delta region of the country and last from less than 3 months to about 11 months respectively. Over 80 per cent of Nigeria's population depends on rain fed agriculture and fishing as their primary occupation leading to high risk of food production system being adversely affected by the variability in timing and amount of rainfall (Ebele 2016).

According to Olagunju (2015) "desertification caused loss of biodiversity, contributed to disease burden, altered geo-chemical composition of the soil, contribute to water scarcity, reduced agricultural yield, hence contribute to food insecurity, reduced economic growth among other unfavorable impacts"

The Nigerian government has put in place various policies, programmes and projects at combating desertification. These include the EU funded North-East Arid Zone Development Programme, the States' Agricultural Development Programmes, the annual tree planting campaigns in the states mostly affected and the Great Green Wall programme.

This paper attempts to review the main causes of desertification in Northern Nigeria, its adverse effects on the socio-economic life of its people and the various ways put in place by the government and other development partners to control and mitigate its effects. Being a review paper, the methodology adopted for the paper is purely review of existing literature and consulting relevant agencies of government on their efforts at addressing the subject.

## Desertification

Desertification is the most serious environmental problem facing the northern Nigeria. Desertification is associated with loss of the lands biological productivity. Deserts are extremely dry areas with sparse vegetation. According to (Terkula, 2009) Nigeria losses about 350,000 hectares of land every year to desert encroachment and most of this land is in the frontline states in the northern Nigeria. Farmers are constantly faced with inadequate rainfall, the level of food crop and livestock productions in these areas have degenerated. Desertification can broadly be classified in three (3) categories of Mild Desertification where land productivity drops to between 10-25 per cent, Serious Desertification productivity drops to 25-50 per cent and Very Serious Desertification is a drop in land productivity to 50 per cent and above which leads to serious gullies and sand dunes (Mohammed, 2015).

### Features of a Desertification Process

(OLAGUNJU, 2015) noted the following as the features of desertification:

★ Impoverishment of vegetative cover

  ★ Reduced quantity, available and accessible soil moisture
  ★ Deterioration of the texture, structure, nutrient status of the soil
  ★ Reduced bio-diversity and presence of xeric shrub lands
  ★ Increased soil erosion

## Impacts of Desertification

The consequences of desertification can be seen in the increased food shortages, water scarcity, economic hardship and political unrest. Desertification can also affect ecological balance and provides a favorable environment for diseases and vector-borne diseases. The recent increase in reported cases of meningitis in northern Nigeria may not be unconnected with the increased temperature.

The drying up of Lake Chad has been attributed to the impact of desertification and as reported by (Goni, 2016) the resultant effect of the drying of Lake Chad is decline in fishing activities, decline in irrigation activities/food insecurity, decline in income and deterioration of living conditions, increased conflict over resources and increase crime rate (and restiveness) and increased migration. International media is full of reports of increased migration from Africa to Europe across the Mediterranean Sea.

The persistence of desertification reduces national food production and further increases the chances of food and raw materials imports for industrial growth. This further depletes the foreign exchange and increases job losses.

There are also increased conflicts between nomads and farmers associated with competing needs of scarce resources, overgrazing lands and agricultural lands for crop cultivation. According to the Institute for Peace and Conflict Resolution, Northern Nigeria has witnessed a fair share of conflicts resulting from the effects of desertification, the situation which has resulted in brutal clashes between farmers and cattle herders in Kaduna and Benue states of the northern Nigeria.

The impacts of drought are extremely serious and often dramatic particularly to the most vulnerable groups-women and children because during emergency the men often migrate to the urban areas leaving the women and the children with the responsibility of providing food, water, energy and rearing of the domestic animals.

Various techniques have been used by farmers in the region to preserve soil water. These techniques include Terracing, Mulching, Drip irrigation, *etc.*

## Drought

Drought can be defined as naturally occurring phenomenon though, usually aggravated by human activities that occurs in a particular period in a particular area such that precipitation is significantly below normal recorded levels, causing deterioration of land productive system and variably agricultural outputs (Olagunju, 2015). Droughts are more common in arid and semiarid lands. Persistent droughts and poorly managed land and water resources increase the chances of desertification.

The causes of drought and desertification are the resultant interaction between of climatic factors.

## Climatic Variability

This is a very important factor in the causes of drought and desertification. This phenomenon has been related to increasing release of greenhouse gases that deplete the ozone layer and increase exposure to ultraviolet light leading to global warming. This is further aggravated in Nigeria by gas flaring and activities of the militants in the Niger Delta that burst oil and gas pipelines for political causes.

Some of the alterations caused include:

1. Decrease in the amount of rainfall in dry lands making arid and semi-arid lands dryer.
2. Drying up of water sources due to high temperature.
3. Poor vegetation.

## Human Activity

Human activity contribute to desertification through poor land use such as rapid expansion of agricultural lands, cultivation on marginal soils, overgrazing, tree felling, bush burning, *etc*. Nigeria is considered the world's highest deforested country and has lost about 55.7 per cent of its primary forests (Olagunju, 2015). This activity reduces the amount of Carbon dioxide that is removed from the atmosphere. A lot of damage has been done to Nigeria's land through the processes of deforestation, notably contributing to the overwhelming trend of desertification. Expansion of agricultural land to meet up with the food requirements of the increasing population has led to the degradation of land in northern Nigeria. Overgrazing removes the vegetation cover that protects soil from erosion and degrades natural vegetation that leads to desertification. This is further aggravated by activities of the nomadic Fulani cattle rearers in northern Nigeria. Other anthropogenic factors include urbanization and poor irrigation management.

## Government Measures to Mitigate Drought

Some of the measures taken for the management of drought in Nigeria include:

1. Provisions of emergency relief materials by the National Emergency Relief Agency, States' Emergency Agencies, International Development Partners, *etc*.
2. Provision of feed supplement for livestock.
3. Provisions of irrigation and water pumps.
4. Establishment of range lands of about 20 hectares in each of the most affected local governments.
5. Establishments of National Strategic Grain Reserve, the government of Nigeria established the National Strategic Grain Reserve as one of the coping measures in order to reduce the risk of drought. The scheme has 12 silos across the country with a total capacity of 350,000 MT. Contracts for 20 additional silos to boost the storage capacity to 1,350,000 MT has been awarded. In addition to the silos; there are additional warehouses for

storage of grains. This is with a view to boost national capacity to reduce risk in drought.

6. Creation of Ecological Fund Office under the office of the Secretary to the Government of the Federation. The Fund was created as an intervention facility to address the myriad of ecological problems including drought and desertification ravaging communities in the country.

7. Creation of the Federal Ministry of Environment with a dedicated department on Drought and Desertification Amelioration; and other activities such as sand dunes fixation programme, oasis inventory and rehabilitation, Rangeland establishment, drought forecasting and formulation of drought and desertification policies development of national drought preparedness plan, rain water harvesting, the great green wall programme to halt desert encroachment.

8. The Great Green Wall programme is a Pan-African Initiative to address land degradation and desertification, boost food security and support communities to adapt to climate change in the Sahel-Sahara region of Africa. This is a line of trees that spans the west end to east end of Africa across the Sahara desert and the Sahel savannah region.

## Government Efforts in Desertification Control

The effort of the Nigerian government in desertification control can be traced back to the colonial era when investigation reports by the Anglo-French commission in 1937 directed the emirates in the northern protectorate to embark on tree planting exercise to halt desert encroachment. The Federal Government in 1977 set up the Arid Zone Afforestation Project. This project launched tree planting campaigns and forestry projects to check deforestation. This involved production and distribution of seedlings which led to the establishment of shelter belts along the northern borders of the country (Olagunju, 2015) and the Neem trees (*Azadirachta indica*) lining the streets of major cities in the North.

The Nigerian government also in its National Action Programme (NAP) to combat desertification and ensure sustainable development identified sectorial policies which included the National Environmental Policy, National Agricultural Policy, National Forestry Policy (National Action Plan to Combat Desertification and Mitigate the Effect of Drought in Nigeria, 2005).

The National Agricultural Policy plan in combating drought and desertification include:

1. Protection of Agricultural lands against drought, desertification, soil erosion and flood, protection and conservation of forests, promotion of alternative sources of energy, integrated water resources management and promotion of appropriate farming systems.

2. The Nigerian government also signed the United Nations Conference on Environment and Development (UNCED) and other initiatives which include the establishment of National Coordinating Committee on

Desertification Control which is the national implementing body of the Conference in Nigeria.

3. Other efforts include the River Basin Development programmes and several dams constructed for irrigation and power generation across the country and bridging the gap between policy formulation and implementation strategies by creating government departments and agencies responsible for this.

## Conclusions

The menace of drought and desertification in northern Nigeria and its impact on the socio-economic life of its people led to loss of arable lands and food shortages, conflict between farmers and nomads resulting in violent clashes and loss of lives, ethnic and religious insurgency. The efforts of the government and key stakeholders in addressing the twin challenges have been enumerated in the paper. However, more synergistic effort needs to be put in place to halt the south-ward movement of the Sahara desert and also roll it back.

## Recommendations

1. Strengthen the capacity of the relevant Departments in the Federal Ministry of Environment to coordinate activities for combating desertification and mitigate the impact of drought. This will enhance response effectiveness, adequate preparedness planning and maximize mitigation efforts.

2. Training and capacity building of relevant Governments Ministries, Departments and Agencies for drought monitoring, assessment and forestry

3. Raising awareness at local, regional and global level on the adverse effect of desertification and ways of mitigating its adverse effect. This will serve as a preventive measure against the anthropogenic factors causing desertification.

4. Best agricultural practice on marginal soils and improved irrigation techniques

5. Continues tree planting campaign and planting of economic trees in affected areas for improved livelihood of rural dwellers (such as Gum Arabic, Shea nut, neem, date palm, *etc.*)

6. Government to put measures in place to support ranch system as against nomadism. This has the advantages of saving lives from continuous clashes between farmers and pastoralists, increased agricultural production and food security.

7. Development of sustainable alternative energy sources such as solar energy, wind energy, bio fuels and use of efficient cooking systems.

8. Strong political will led by the executive, supported by the legislature and the judiciary.

9. Community based action programs on drought management and desertification control

10. Bridging the gap between policy formulation and implementation strategies.

# References

1. Abdulkadir, A. (2016). Water as a Vital Resource and a Boost of Regional Economy. Paper presented at Regional Workshop of Capacity Building of Technical Experts and Stakeholders on Sustainable Development of ECOWAS Region: Water Resources and Sustainable Environmental Management in the ECOWAS Region Held at University of Maiduguri, Nigeria. 31$^{st}$ October – 2$^{nd}$ November 2016.

2. About the Sahel http: //www.unocha.org/sahel/about-sahel Retrieved 4/24/2017

3. Drought Conditions and Management Strategies in Nigeria. http: //www.droughtmanagement.info/wp.content/uploads/2016/10/ws6-Nigeria_EN.pdf.

4. Ebele, E. N. and Emodi, N. V. (2016). Climate Change and its impact in Nigerian Economy Journal of Scientific Research and Reports. 10(6): pp. 1-13, 2016: Article no. JSRR.25162

5. Goni, I. (2016). Mitigating Environmental Degradation and Water Pollution Emanating from Industrialisation, Urbanisation, (Water ways Blockage) Silting, Dumping of Waste, Oil Spillage, Gas Flaring, Carbon Emission, *etc.*). Paper presented at Regional Workshop of Capacity Building of Technical Experts and Stakeholders on Sustainable Development of ECOWAS Region: Water Resources and Sustainable Environmental Management in the ECOWAS Region Held at University of Maiduguri, Nigeria. 31$^{st}$ October – 2$^{nd}$ November 2016.

6. Hiroshi, K. Combating Desertification and Drought. Encyclopedia of Life Support Systems (EOLSS) (UNESCO-EOLSS Sample Chapters) Retrieved 4/24/2017.

7. Medugu N.I. Majid, M.R. and Johar, F. (2011). Drought and Desertification Management in Arid and Semi Arid Zones of Northern Nigeria. http: //www.researchgate.net/publication/228462893 Retrieved 4/24/2017.

8. Mohammed N.T. (2015). Desertification in Northern Nigeria: Causes and Implications for Food Security Peak Journal of Social Sciences and Humanities Vol. 3 (2) pp 22 – 31, March 2015.

9. Olagunju, T.E. (2015). Drought, Desertification and the Nigerian Environment: A Review Journal of Ecology and Natural Environment, Vol. 7(7) pp 196-209, July, 2015

10. Oil Production www.nnpcgroup.com Retrieved 17/05/2017.

11. Political Map of Nigeria- https: //www.ezilon.com/maps/africa/nigeria-maps.html Retrieved 12/22/2017.

*Chapter 12*

# Prevalence of *Schistosoma haematobium* in Sudan Dry Land: A Case Study at Al-Rahad City, North Kordofan State

*Abdel-Moneim Mohamed Salim*

*Department of Biology, Faculty of Science,*
*Taif University (Turaba Branch),*
*21995, Kingdom of Saudi Arabia*
*E-mail: salimabdelmoneim55@msn.com*

## Abstract

In Sudan dry land, the local communities depend on natural resources for alternative. They parctise subsistence agriculture and animal herding. Some of these activities may affect their life negatively by increasing the probability of getting infectious diseases such as Schistosomiasis. This disease is amongst the areas projecting tremendously higher prevalence and intensity especially during childhood. A cross sectional study was conducted to determine the prevalence and intensity of urogenital schistosomiasis among primary school children in Al - Rahad city in North Kordofan State, Sudan. A total of 114 students from 5 primary schools were chosen for sampling. They were classified into 3 age groups:7-9,10-12 and 13-15 years old. Their urine samples were examined using filtration laboratory technique for the infection with *S. haematobium* using the standard filtration technique. Data were analyzed statistically using SPSS, with respect to *S. haematobium* prevalence related to child sex and age group.

The study has recorded an overall higher prevalence *S. haematobium* among male students, (54.4 per cent), as compared to females, (45.6 per cent). Results of the Student t-test insignificant difference of egg count between the two sexes, (p= 0.467). ANOVA results indicated significant difference of the egg count between and within the age groups, (p= P=0.06). Results of Pearson Chi-square test revealed high correlation of *S. haematobium* infection to age group of infected children. These findings indicated that Al-Rahad city is endemic to urogenital schistosomiasis especially among males as compared to females and therefore, in order to mitigate the high intensity of infections, a control program is required.

The objectives of this study are to examine the prevalence Schistosomiasis in Sudan dry land especially in Kordofan State and to evaluate the infection rate among school children in Al-Rahd city. It also, meant to through some light on the endemic diseases in the area.

*Keywords:* Sudan, Dryland Schistosoma haematobium, Al-Rahad, Filtration technique, Urogenital schistosomiasis.

## Introduction

Dry land ecosystems are very fragile, and can rarely sustain the increased pressures that result from intense population growth. In Sudan, many of these areas are inappropriately opened to development Schistosomiasis (Bilharziasis or Bilharzia) is a chronic and acute parasitic disease caused by digenetic blood vessel-dwelling trematode of the genus *Schistosoma*. It occurs in the intestinal and urogenital forms. Urogenital schistosomiasis caused by *Schistosoma haematobium* (*S. haematobium*) and is characterized by progressive damage to the bladder, ureters and kidneys. In 2010 the disease has been enlisted as one of the Neglected Tropical Diseases (NTDs) that present some of the most universal health problems in the world. Around 119 million people worldwide while the regions surrounding Sub-Saharan Africa remains the area of prevalence with an overall mortality rate estimated to be at least 2 per 1,000 infected patients per year (Melchers *et al.*, 2014). This has led urinary schistosomiasis to be deadliest parasitic disease in the NTDs (WHO,2010, Hotez *et al.*, 2006). In fact, only malaria accounts for more diseases 48 than schistosomiasis.

Infection occurs by contact with stagnant or slow-moving freshwater where infected *Bulinus* snails live. Preferably, lakes, natural streams, and ponds submerged with infected *Bulinus* snails forms the typical sources for infection. Irrigation systems, dams, and man-made water reservoirs are proven to significantly contribute in increasing the incidence of disease in the last few decades (McManus, and Loukas 2008). Thus, the main risk groups are children under 15 years of age, specific occupational groups (freshwater fishermen, irrigation workers, farmers), and women fetching water for home use and other groups using infested water for domestic uses (WHO,2002). Patterns of sanitation, water supply, and human water use are crucial elements in making people vulnerable to infection. Moreover, Play habits of school-aged children such as swimming or fishing in infested water make them especially vulnerable to infection. According to Nokes *et al.* (1999), schistosomiasis is considered a significant risk factor in children because it result in chronic anemia, malnutrition, growth stunting, protein calorie, cognitive disability, and poor school performance. Schistosomiasis decrease in sensitivity in adulthood and prevalence rises to a peak during the years 10–15, then declines through the 20s, 30s and 40s to well less than half of the childhood peak, due to behvioural activity shown in less contact with water and immunity to infection. Similar observations were reported by Nmorsi *et al.* (2007).

In Sudan, urogenital schistosomiasis is widespread and constitutes a critical public health problem mainly among children in their school going age as compared to other age groups (Ahmed, *et al.*, 2012 and Seghor, *et al.*, 2014). A number of factors have been deemed responsible because of higher rates between the specified age groups. Thus, for instance, include the increased water activities such as fishing or swimming (Agrawal, and Rao, 2011). Besides, improper hygiene arrangements provide better opportunities for the spread of infection, (WHO,2002). Some other factors include blood vessels supplying genitourinary system as well as immunological factors (Mutpi *et al.*, 2008, and Sam-Wobo, *et al.*, 2011). Furthermore, people in developing countries such as Sudan have poor living standards and healthcare facilities which are needed for proper control of the disease (Monde, *et al.*, 2015).

Southern Kordofan State lies on the border between Sudan and South Sudan. For more than 20 years, this state was the scene for war between Sudan and South Sudan that drastically affected the social and economic status of the 1.6 million people living in the State. People depend on subsistence agriculture and herding which may intensify infection since the disease associated with poverty and poor living conditions especially among risk group (WHO,1993 and Elawad 2005). Water is provided mainly from dams, wells, superficial rainwater collections and water pumps. Many water ponds are present and stay wet year-round. The availability of latrines in SK State is very low (less than 20 per cent), and the use of available latrines needs to be improved, (Federal Ministry of Health-Sudan, (2007).)There have been two major reasons for focusing the study in this area of Sudan. Firstly, it has been referred as the most vulnerable areas in the continent of Africa, where a profusion of schistosomiasis has been growing at unprecedented rates (Abou-Zeid, et al., 2012, and Dahab, and El-Bingawi, 2012). The communities living there are found to be economically stressed and mostly dwell nearby rivers or water streams (Afifi, *et al.*, 2016). Secondly, there have been only a few studies reported regarding detection of schistosomiasis among the given population (Elfaki *et al.*, 2016). Moreover the disease is responsible for extensive morbidity and mortality in up to 90 per cent of severe infections detected in children of the state and thus recognized as one of 10 tropical diseases of most concern to the World Health Organization, (Amin, and Satti 1973, WHO, 2010 and Amin *et al.*, 2012).

The study, therefore aims to bridge the research gap and provides empirical evidences for formulating control programs for urinary schistosomiasis among the inhabitants of the city in accordance to the World Health Organization recommendations.(WHO).

## Materials and Methods

### Study Area

The study focused on the area of Al-Rahad city (12°43'0" N, 30°39'0" E, 85 1608 ft above the sea level), located in North Kordofan State, 545 km west of Khartoum, the capital of Sudan, North-East Africa (Figure 12.1). It is a junction station on the western line of the national railway network in Sudan and its population amounts to 26,273 according to the last census 2009.

**Figure 12.1: Location Map of Al-Rahd City.**

The climate is Semi-desert to subtropical savannah, with average annual rainfall of 400-600 mm. The rainy season lasts for about five months, usually begins in May and reaches its peak in August and continues up until September. One characteristic geographical feature of the area is the absence of rivers. Alternatively; at the west of the Al-Rahad city, there is a huge freshwater body (locally known as Al-Turda) that traverses the locality. Al-Turda is a permanent man-made lake used to store rainwater; its annual storage capacity is 52,000,000 Cubic Meters that are used to secure the City, the 98 villages, and the local farms in the surroundings against the annual 99 threatening of "Lack of Water" during the critical summer months. 100 Therefore, people in this part of Sudan depend mainly on Al-Turda for 101 drinking, domestic, fishing, animal and agricultural purposes and other water needs.

Throughout many visits to the study area, most of the children were observed to play and swim in Al-Turda after school and in their leisure time. As a result, contact with the water is an absolute necessity for the majority of them, besides that, their living conditions and individual hygienic practices were also poor. So, the populations of the selected schools were representative of the parasitological situation in this area.

**Study Population and Subject Selection**

This is a cross-sectional study carried out among school children aged 7-13 years old in Al-Rahad city during three weeks in September 2014. School children were randomly selected from five primary schools in the city. Data were collected with exclusive reference codes given to each participant. Each child was administered a simple health structured questionnaire to collect social and economic data of the pupils. Information on age, sex, educational background perception, knowledge of symptoms, sources of water supply/mode of transmission and health implications of schistosomiasis were obtained. The class teachers administered the questionnaire to the children in the language (Yoruba or English) he/she understand best.

## Ethical Approval

The study received ethical clearance from the State government after it, a pre-survey visit to the study area was made during which consultations and discussions were held with local governors, the local community leaders and primary school teachers who were informed about the aims and methods of the study. Inormal seminars were conducted to raise public awareness before sampling commenced.

## *Informed Consent*

Verbal informed consent of the school children was obtained after the headmaster of the school had explained the project to the pupils. They were allowed a day to discuss and obtain permission of their parents. Children whose parents or legal guardians objected to their participation were excluded. On the sample collection day the class teacher helped to confirm the consent from the pupils. Moreover, all those who participated in the study were informed that they could withdraw without any penalties, if they were not comfortable with participating. The children who took medication for schistosomiasis three weeks prior to and during the data collection and children who were seriously ill during data collection were also excluded.

## *Collection of Urine Samples*

Samples were collected in sterile plastic containers each assigned to an individual specific code for the sake of reference and record. The containers had a wide mouth and showed a capacity of 200 ml marked well with the identification numbers. Children were given the autonomy to self-collect their samples. However, they were given brief instructions for careful collection of the samples. They were instructed to collect urine by the mark of 10 ml and tightly cover bottles with the cap. Samples were collected in schools during the day timings precisely 10:00 am to 2:00 pm due to the circadian pattern of egg excretion, *i.e.* the excretion of eggs is at its peak and may bring expectedly accurate results (Senghor *et al.*, 2014). The appearance of urine was noted for each sample. Formal saline (3 drops) was added to preserve the samples and avoid hatching of eggs (Agrawal,2012). The samples collected were 150 immediately stored in the black cellophane bags to avoid contact with the sunlight and were immediately moved to the field laboratory for examination could be performed then and there.

## Parasitological Diagnosis

At the field laboratory, about 10 ml of an individual's urine were drawn into a plastic syringe and then discharged through polycarbonate (Nucleopore®) filter (25 mm diameter, 12.0 μm pores) according to the method described previously (Kean *et al.*, 1978). The specimens (the filters) were microscopically examined for S. haematobium eggs using the 10 x and 40 x objectives. The examination and egg count was repeated on another portion of the sample. Results expressions were formed through presenting the mean number of eggs for every 10 ml of urine, if more than 50 eggs are present; there is no need to continue the counting. The cases of schistosomiasis were defined as children with at least one *S. haematobium* egg on microscopic examination of urine. Categorization of the intensity of infection was done according to instructions by WHO (2002), *i.e.* it is based on the egg count contained per 10 ml of urine. These include the non-infected samples (with no eggs), low intensity samples (containing 50 or less eggs per 10 ml), and highly infected samples (with 50 or more eggs per 10 ml). Individuals who tested positive for schistosomiasis infection were treated with praziquantel (Shin Poong, Seoul, South Korea) according to WHO guidelines.

### Data Analysis

The Statistical Package for Social Science, (SPSS) was used to analyze the data. The relationships between the characteristics of infection (prevalence and intensity). The chi-square test was used to test for the vaiance association. ANOVA and the Student t-test, were used to analyze prevelency and egg count between sex and ade group of infected children.

## Results

### Sample Characteristics

Classification of sampled groups indicated that 45.6 per cent are between 10-12, 28.1 per cent is between 13-15 years, and 26.3 per cent is between 7-9 (Figure 12.2).

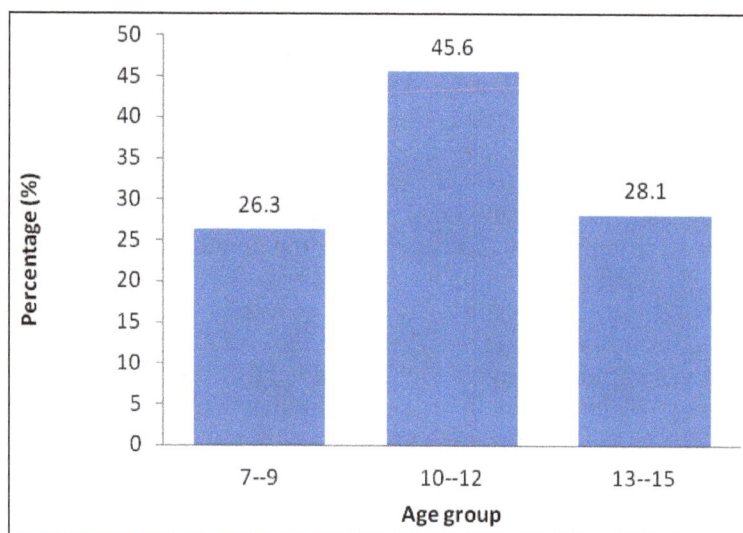

**Figure 12.2: Percentage of the different Age Group Included in the Study at Al-Rahad City during 2014.**

### Infection

Incidence of *S. haematobium* infection in Al-Rahad city is evaluated in relation to age group of the sampled children. Results obtained illustrated that 45.6 per cent of the encountered infected cases is reported age group 10-12 years. Whereas,28.1 and 26.3 per cent of cases are recorded within the age range13-15 and 7-9 years, respectivly. Gender wise male are shown to get more infection of (54.4 per cent) as compared to females (45.6 per cent) (Figure 12.3).

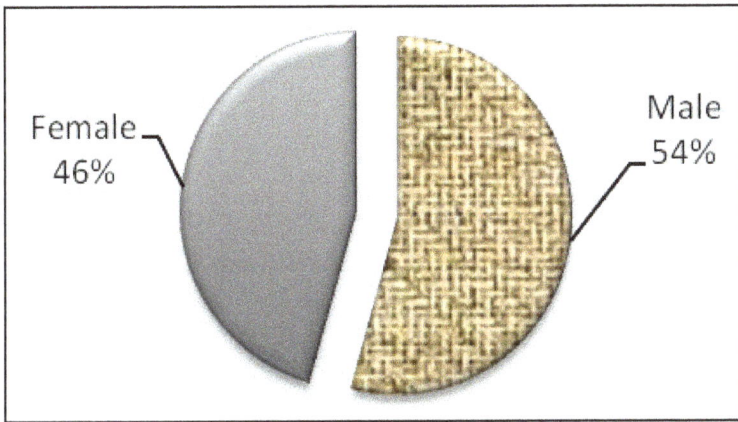

**Figure 12.3: Prevalence of *S. haematobium* According to the Infected Child Sex at Al-Rahad City during 2014.**

## *S. haematobium*'s Egg Secretion

The count *of the S. haematobium's* eggs secreted in urine samples of the infected children shows significant variations in prevalence of relation to sex. Insignificant variation in intensity of egg secreted in male samples compared to females (Table 12.1).

**Table 12.1: Value of the Student t-test as Calculated for Egg Secretion in Relation to the Sex of the Infected Child at Al-Rahad City during 2014**

| Sex | Means | P – value |
| --- | --- | --- |
| Males | 8.9469 | 0.206 ns |
| Females | 9.4533 | |

* Significant at P=0.05.

## The Impact of Age Group on *S. haematobium* Egg Count

*S. haematobium* egg **count** in relation to the age group of infected children was studied. ANOVA results revealed considerable differences between and within groups. It indicated that the intensity of *S. haematobium* is influenced by both sex and age group. However, the impact of age groups on egg count is insignificant, (p= 0.467). A significant difference is found in prevalence between egg count within the age group, (P=0.06) (Table 12.2). Likewise, *S. haematobium* infection was correlated with age group of infected children. Results of Pearson Chi-square test revealed high correlation values as shown in Table 12.3.

**Table 12.2: ANOVA Table for Impact Infected Children Age Group on**
*S. haematobium* **Egg Count at Al-Rahad City during 2014**

| Chi-Square Tests | Value | Df | Asymp. Sig. (2-sided) |
|---|---|---|---|
| Pearson Chi-Square | 7.748 a | 2 | 0.021 |
| Likelihood Ratio | 7.926 | 2 | 0.019 |
| Linear-by-Linear Association | 7.678 | 1 | 0.006 |
| N of Valid Cases | 114 | | |

**Table 12.3: Pearson Chi-square Test for** *S. haematobium* **Infection and Age Group at Al-Rahad City during 2014**

| Egg count | Sum of Squares | df | Mean Square | F | Sig. |
|---|---|---|---|---|---|
| Between Groups | 4.964 | 2 | 2.482 | 0.766 | 0.467ns |
| Within Groups | 359.827 | 111 | 3.242 | | 0.06* |
| Total | 364.791 | 113 | | | |

* Significant at P=0.05.

## Discussion

The study recorded an overall prevalence of *S. haematobium* infection from selected age grouped children at Al-Rahad city during 2014. It was found that indicated that 45.6 per cent of the sampled cases are from 7-9 years old. Also, infected males exceeded female ones. These results were in conformity of earlier studies conducted by (Senghor, *et al.*, 2014; Dahab, *et al.*, 2012; Liao, *et al.*, 2011; Abou-Zeid, *et al.*, 2013 and Deribe, *et al.*, 2011). These authors claimed that impacts of child age in relation to susceptibility of infection with *S. haematobium*. On the other hand, these findings disagreed with that of Ekpo, *et al.* (2010), who indicated insignificant differences in intensity of infection with the child age. This could be as ascribed to the impact of environmental hygiene condition that might exist among the study groups. Similar observations were made by Naphtali, *et al.* (2017),who correlated infection with *S. haematobium* to primary school children age group and parent education. Elias, *et al.* (1994), formerly observed increased prevelance of *S. haematobium within* younger age group in a study in the Rahad irrigated scheme, Sudan.

This study also revealed that males had higher prevalence rates of infection than the females. This agreed with Shashie, *et al.* (2015), who indicated that males showed the highest infection intensity of infection than females. They attributed their findings to the fact that . male children have more affinity to be infected than females because they are engaged in water contact activities that necessitated them to be in direct contact with water bodies. These activities include farming,fishing, bathing, and watering cattle. The results are in line with findings of (Risikat, and

Ayoade,2012), but in contraction with those of Gyuse *et al.* (2010). And Dabo, *et al.* (2010).

The count *of the S. haematobium's* eggs secreted in urine samples of the infected children shows significant variations in egg secretion of relation to sex. Insignificant variation in intensity of egg secreted in male samples compared to females. These results could be ascribed to the differences in the male and female intensities of infection, which may reflect the degree of sensitization and host response to the invading parasites as well as the extent of worm burden in the individual subject.

## Conclusions

In Sudan the risk for *S. haematobium* is widespread in the different regions and school age children were at a higher risk of *S. haematobium* infection than the other age groups. Due to many factors such as higher rates of water activities, anatomical vasculature supplying genitourinary structures and immunological factors, school-aged children are the group at highest risk of contracting *S. haematobium* infection, Mohammed,(2006).

A field survey was carried out to study prevalence schistosomiasis in Sudan dry land with respect to the incidence among school children in Al-Rahd city. A higher prevalence of the infection was detected among male children than females. Also, higher prevalence was observed among children aged 10-12. A higher egg activity occurred among male children as compared to females. It is apparent that one way of reducing the disease prevalence is through minimizing the contact of the children with the infected water.sources. Also, treatment of children with higher intensity of schistosomiasis may be applied through large-scale chemotherapy. Provision of a piped water the residential area is a prerequisite to prevent childern from contracting the infection.

## References

1. Abou-Zeid AH, Abkar TA and Mohamed RO. (2013). Schistosomiasis infection among primary school students in a war zone, Southern Kordofan State, Sudan: a cross-sectional study. *BMC Public Health*, 13: 643-650.

2. Abou-Zeid AHA, Abkar TA, Mohamed RO. (2012). Schistosomiasis and soil-transmitted helminths among an adult population in a war affected area, Southern Kordofan state, Sudan. *Parasite Vectors*, 5: 133-147.

3. Afifi A, Ahmed AAA, Sulieman Y and Pengsakul T. (2016). Epidemiology of Schistosomiasis among Villagers of the New Halfa Agricultural Scheme, Sudan. *Iran J. Parasit.*, 11(1): 110-115.

4. Agrawal MC, Rao VG. (2011). Indian Schistosomes: A Need for Further Investigations. *J. Parasitol Res.* 250868-250877.

5. Agrawal MC. (2012). Schistosomes and Schistosomiasis in South Asia. Springer India; pp. 187-213.

6. Ahmed AM, Abbas H, Mansour AF, Gasim G, Adam I. (2012). Schistosoma haematobium infections among school children in central Sudan one year after treatment with praziquantel. *Parasit Vectors*, 5: 108-117.

7.  Amin, M. and Satti, M. 1973. A General Review on Schistosomiasis in the Sudan. *Am J Trop Med Hyg* 60(4): 556-65.

8.  Amin, M., Swar, M., Kardaman, M., Elhussein, A., Nouman, G., Mahmoud, A., Appiah, A. Babiker, A. and Homdeida, M. 2012.Treatment of pre-school children under 6 years of age for schistosomiasis: safety, efficacy and acceptability of praziquantel. *Journal of Medical Sciences,* .7: 67-76

9.  Dabo A, Badawi HM, Bary B. and Doumbo OK. (2011). Urinary schistosomiasis among preschool-aged children in Sahelian rural communities in Mali. *Parasit. Vectors,* 4: 21-27.

10. Dahab TO and El-Bingawi HM. (2012). Epidemiological survey: *Schistosoma haematobium* in schoolchildren of White Nile areas, Khartoum.

11. Deribe K, Eldaw A, Hadziabduli S, Kailie E, Omer MD, Mohammed AE. (2011). High prevalence of urinary schistosomiasis in two communities in South Darfur: implication for interventions. *Parasit Vectors,* 4: 14-23.

12. Ekpo UF, Laja-Deile A, Oluwole AS, Sam-Wobo SO, Mafiana CF. (2010)

13. Elawad,K.H. (2005). The Prevalence of Schistosomiasis among primary School Children in Barakat area- Gezira State. M.Sc. University of Khartoum Faculty of Public and Environmental Health, Department of Epidemiology

14. Elfaki TEM, Arndts K, Wiszniewsky A, Ritter M, Goreish IA., Misk El Yemen A. (2016). Multivariable Regression Analysis in Schistosoma mansoni- Infected Individuals in the Sudan Reveals Unique Immunoepidemiological Profiles in Uninfected, egg+ and Non-egg+ Infected Individuals. *PLOS Negl Trop Dis,* 10(5): e0004629-0004637.

15. Elias,E., Daffalla,A.,Lassen, JM., Madsen,H. and Christensen, NO. (1992). *Schistosoma haematobium* infection patterns in the Rahad Irrigation Scheme, Sudan. *ACTA Tro.,* 58(2): 115-25.

16. Federal Ministry of Health-Sudan (2007). The Sudan Household Health Survey. Report from the Federal Ministry of Health., 106-121. Available from: http: // reliefweb.int / report / sudan / sudan-household-and-health-survey-second-round-2010-summary-report

17. Gyuse KI, Ofoezie IE and Ogunniyi T. (2010). The effect of urinary schistosomiasis on the health of children in selected rural communities of Osun State, Nigeria. *J Trop Med Parasitol,* 33: 7-16.

18. Hotez PJ, Savioli L, Fenwick A.(2006). Neglected Tropical Diseases of the Middle East and North Africa: Review of Their Prevalence, Distribution, and Opportunities for Control. *PLOS Negl Trop Dis* 2012; 6 (2), e 1475. (4) Gryseels B, Polman K, Clerinx J, Kestens L. Human schistosomiasis. *Lancet,* 368: 1106–1118.

19. Kean BH, Mott KE and Russell AJ. (1978). Tropical medicine and parasitology classic investigations. Cornell University Press, Ithaca, N.Y; pp. 1188-1191.

20. Liao. CW, Sukati H, Nara T, Tsubouchi A, Chou CM, Jian JY. (2011). Prevalence of *Schistosoma haematobium* infection among schoolchildren in remote areas devoid of sanitation in northwestern Swaziland, Southern Africa. *Jpn. J. Infect. Dis*, 64(4): 322-326.

21. McManus DP and Loukas A. (2008). Current status of vaccines for schistosomiasis. *Clinical Microbiol Rev.*, 21: 225–242.

22. Melchers NVV, van Dam GJ, Shaproski D, Kahama AI, Brienen EA, Vennervald BJ. (2014). Diagnostic performance of Schistosoma real-time PCR in urine samples from Kenyan children infected with *Schistosoma haematobium*: day-to-day variation and follow-up after praziquantel treatment. *PLOS Negl Trop Dis.* 8(4), e 0002807.

23. Mohammed EH, Eltayeb M, and Ibrahim H. (2006). Haematological and biochemical morbidity of Schistosoma haematobium in school children in Sudan. *Sultan Qaboos Univ Med J.*, 6: 59-64.

24. Monde C, Syampungani S, van den Brink P. (2015). Exploring the potential of host-environment relationship in the control of schistosomiasis in Africa. *Afric J. Aq. Sc.* 40: 47–55.

25. Mutapi F, Burchmore R, Mduluza T, Midzi N, Turner CM, Maizels RM. (2008). Age related and infection intensity-related shifts in antibody recognition of defined protein antigens in a schistosome-exposed population. *J Infect Dis.* 198: 167–175

26. Naphtali, R., Yaro, M. and Arubi, M. (2017). Prevalence of Schistosoma haematobium among Primary School Children in Girei Local Government Area, Adamawa State, Nigeria. *IOSR Journal of Nursing and Health Science (IOSR-JNHS)* e-ISSN: 2320–1959.p- ISSN: 2320–1940 Volume 6, Issue 1 Ver. II (Jan.-Feb. 2017), PP 48-50 www.iosrjournals.org

27. See comment in PubMed Commons belowNmorsi OP1, Kwandu UN, Ebiaguanye LM (2007). *Schistosoma haematobium* and urinary tract pathogens co-infections in a rural community of Edo State, Nigeria. 39(2): 85-90.

28. Nokes C, McGarvey ST, Shiue L, Wu G, Wu H, Bundy DA, Olds GR. (1999). Evidence for an improvement in cognitive function following treatment of *Schistosoma japonicum* infection in Chinese primary schoolchildren of Osun State, Nigeria. *J Trop Med Parasitol.*, 33: 7-16.

29. Parasitology: classic investigations. Cornell University Press, Ithaca, N.Y; (1978). pp. 1188-1191.

30. Risikat SA and Ayoade AA. (2012). Correlation analysis between the prevalence of *Schistosoma haematobium* and water conditions: A Case Study among the School Pupils in Southwestern Nigeria. *IJRRAS*, 13(1): 160-165.

31. Sam-Wobo SO, Idowu JM, Adeleke MA. (2011). Urinary schistosomiasis among children and teenagers near Oyan dam, Abeokuta, Nigeria. *J Rural Trop Public Health*, 10: 57-60.

32. Schistosomiasis on the health of children in selected rural communities

33. Senghor B, Diallo A, Sylla SN, Doucouré S, Ndiath MO, Gaayeb L. (2012). Prevalence and intensity of urinary schistosomiasis among school children in the district of Niakhar, region of Fatick, Senegal. *Parasit Vectors* 2014; 7: 5-13. *Sudan Med J* 48(2): 135-140.

34. Shashie, G., agersew, A., Sisay, G., Zeleke, M. and Berhanu, E. (2015). Prevalence of urinary schistosomiasis and associated risk factors among Abobo Primary School children in Gambella regional State, Southwestern Ethiopia: a cross sectional study. *Journal of Parasites and Vectors*. 215 (8): 822-825.

35. *Sudan Medical Journal*, 11: 86-91.

36. WHO (1993). "The control of schistosomiasis", technique report series, vol. 71, No. 820, pp. 206-213, Geneva.

37. WHO (2010). Climate Change and Health. Media Center, fact sheet.

38. WHO (2010). Working to overcome the global impact of neglected tropical diseases: first WHO report on neglected tropical diseases. Geneva: World Health Organization, 129–134.

39. World Health Organization (2002). Prevention and control of schistosomiasis and soil-transmitted helminthiasis: report of a WHO expert committee. *WHO Tech. Rep. Ser.*, 912: 1-57.

*Chapter 13*

# Sudan Efforts to Combat Desertification: Desertification Research Institute (DRI) Experience

*Maha Ali Abdelatif and Ola Elias Ahmed*

*National Centre for Research (NCR),*
*P.O. Box 2404, People Hall Khartoum, 11113, Sudan*
*E-mail: mahaaali@hotmail.com, olaelias2003@yahoo.com*

## Abstract

In Sudan, awareness about desertification dates back to the 1930s. Since then several attempts and sometimes serious efforts were done to combat desertification. Convinced by environmental, Sudan became one of the first countries to sign and ratify the UNCCD and develop its Desertification National Action Plan aiming to address desertification from a purely domestic perspective. In 1977, Sudan prepared a detailed document, namely "Sudan's Desert Encroachment Control and Rehabilitation Progrm", which was presented to the UN Conference on Desertification. Following that conference, Sudan established the National Desertification Control and Monitoring Unit (NDDU), in 1978. A parallel effort was carried out by the NGOs, National Coordinating Committee for Desertification (NCCD), that coordinated their efforts, especially in the area of raising awareness about the UNCCD. Little success achieved has been attributed to many reasons; among these is lack of scientific research. To fill this gap the Desertification Research Institute (DRI) and Rawakeeb Dryland Research Station (RDLRS) were established. Since its establishment in 1992, the station has been undertaking basic as well as applied research and in the process contributing national and international efforts. From the large number of research projects conducted, 139 refereed papers were published. In addition to that, large numbers of technical papers were presented by DRI researchers in different scientific arenas.

The objective of the present paper is to outline and discuss the outcome of research projects conducted, successes achieved and constraints that hindered the research activities at RDLRS and suggests more support to research in dry lands.

*Keywords: Sudan, Desertification, Desertification control.*

## Introduction

Historically, the world's great deserts have been formed by natural processes interacting over long intervals of time. During most of these times, deserts have grown and shrunk independent of human activities. Palo-deserts are large sand seas now inactive because they are stabilized by vegetation, some extending beyond the present margins of core deserts, such as the Sahara, the largest hot desert (U S Geological Survey, 1999). Desertification has played a significant role in human history, contributing to the collapse of several large empires, such as Carthage, Greece, and the Roman Empire, as well as causing displacement of local populations (Whitford, 2002). The United Nations Convention to Combat Desertification defines the term desertification as 'land degradation in arid, semi-arid and sub-humid areas resulting from various factors including climatic variations and human activities' (UNCCD,1992). Desertification is a dynamic process that is observed in dry and fragile ecosystems. It affects terrestrial areas (topsoil, earth, groundwater reserves, surface run-off), animal and plant populations, as well as human settlements and their amenities, as for instance the terraces and dams (Stringer, 2008).

Drylands occupy approximately 40-41 per cent of Earth's land area,Bauer (2007) and are home to more than 2 billion people. It has been estimated that some 10–20 per cent of drylands are already degraded, the total area affected by desertification being between 6 and 12 million square kilometres, that about 1–6 per cent of the inhabitants can also of drylands live in desertified areas, and that a billion people are under threat from further desertification, World Bank (2009).

Dryland ecosystems are already very fragile, and can rarely sustain the increased pressures that result from intense population growth. Many of these areas are inappropriately opened to development, when they cannot sustain human settlements, Longjun and Xiaohui (2010). The most common cause of desertification is the over cultivation of desert lands, Mares, (1999) Over-cultivation causes the nutrients in the soil to be depleted faster than they are restored. Improper irrigation practices result in salinated soils, and depletion of aquifers, Longjun and Xiaohui (2010). Vegetation plays a major role in determining the biological composition of the soil.

Studies have shown that, in many environments, the rate of erosion and runoff decreases exponentially with increased vegetation cover. Overgrazing removes this vegetation causing erosion and loss of topsoil, Geeson, (2002). At least 90 per cent of the inhabitants of drylands live in developing nations, where they also suffer from poor economic and social condition. This situation is exacerbated by land degradation because of the reduction in productivity, the precariousness of living conditions and the difficulty of access to resources and opportunities, Dobie (2001).

A downward spiral is created in many underdeveloped countries by overgrazing, land exhaustion and over drafting of groundwater in many of the marginally productive world regions due to overpopulation pressures to exploit marginal drylands for farming. Decision-makers are understandably averse to invest in arid zones with low potential. This absence of investment contributes to the marginalization of these zones. When unfavourable agro-climatic conditions are combined with an absence of infrastructure and access to markets, as well as poorly adapted production techniques and an underfed and undereducated population, most such zones are excluded from development, Cornet (2002). Desertification often causes rural lands to become unable to support the same sized populations that previously lived there. This results in mass migrations out of rural areas and into urban areas. These migrations into the cities, often cause large numbers of unemployed people who end up living in slums, (Pasternak, and Schlissel,2001).

Techniques exist for mitigating or reversing the effects of desertification, however there are numerous barriers to their implementation. One of these is that the costs of adopting sustainable agricultural practices sometimes exceed the benefits for individual farmers, even while they are socially and environmentally beneficial. Another issue is a lack of political will, and lack of funding to support land reclamation and anti-desertification programs. Briassoulis,(2005). Desertification is recognized as a major threat to biodiversity. Some countries have developed Biodiversity Action Plans to counter its effects, particularly in relation to the protection of endangered flora and fauna. Reforestation gets at one of the root causes of desertification and isn't just a treatment of the symptoms.

Environmental organizations work in places where deforestation and desertification are contributing to extreme poverty. They focus primarily on educating the local population about the dangers of deforestation and sometimes employ them to grow seedlings, which they transfer to severely deforested areas during the rainy season (Parrillo,(2008).Techniques focus on two aspects: provisioning of water, and fixation and hyper-fertilizing soil. Fixating the soil is often done through the use of shelter belts, woodlots and windbreaks. Windbreaks are made from trees and bushes and are used to reduce soil erosion and evapotranspiration. They were widely encouraged by development agencies from the middle of the 1980s in the Sahel area of Africa. As there are many different types of deserts, there are also different types of desert reclamation methodologies. Sudan has a long experience in combating desertification, which is the major environmental challenge in the State, and Sudan was one of the first States to sign and ratify the UNCCD.

This is a schematic graphic showing how drylands can be developed in response to changes in key human factors (Figure 13.1). The left side of the Figure shows developments that lead to a downward spiral of desertification. The right side shows developments that can help avoid or reduce desertification. In the latter case, land users respond to stresses by improving their agricultural practices on currently used land. This leads to increased livestock and crop productivity, improved human well-being, and political and economic stability. Both development pathways occur today in various dryland areas.

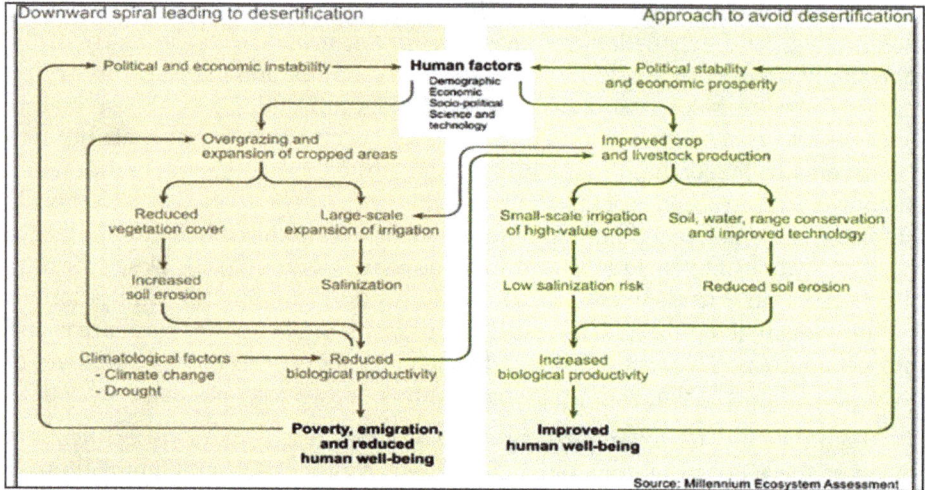

Figure 13.1: Schematic Description of Development Pathways in Drylands (*Source*: Millennium Ecosystem Assessment: A Desertification Synthesis Report, 2005).

In Sudan great efforts have been done to mitigate the effect of desertification but with limited achievements. This could be attributed to lack of scientific research in tackling desertification issues. For this the National Centre for Research (NCR), established Desertification Research Institute (DRI). – this was a good attempt at developing the problem, identifying the gaps in interventions that have been made by the Government but it is incomplete. It needs to be expanded upon.

The objectives of the study should come here!!

Rawakeeb Dryland Research Station (RDLRS): - Where does this section belong? Is it in the background or methodology? It fits more in the methodology section but this should come out clearly.

In order to fulfill the above mentioned objectives and functions the National Centre for Research (NCR) established RDLRS), as a research site.

## RDLRS Physiography

### Location

RDLRS lies between "lat. 15- 2 and 15-36 N and lon. 32-0 and 32-10 E" as shown in Figure 13.2.

### Climate

EL-Rawakeeb lies in tropical semi - arid region whose climate is characterized by a short rainy season (July-October) and high evaporation potential. The relative humidity values are low and thus indicate the general aridity of the area. Air temperature fluctuates and shows a marked rise (47°C) in May and drops in August due to incidence of rains. Average soil temperature is 40 C° while average moisture is 12.5 per cent (El Hag *et al.,* 2006).

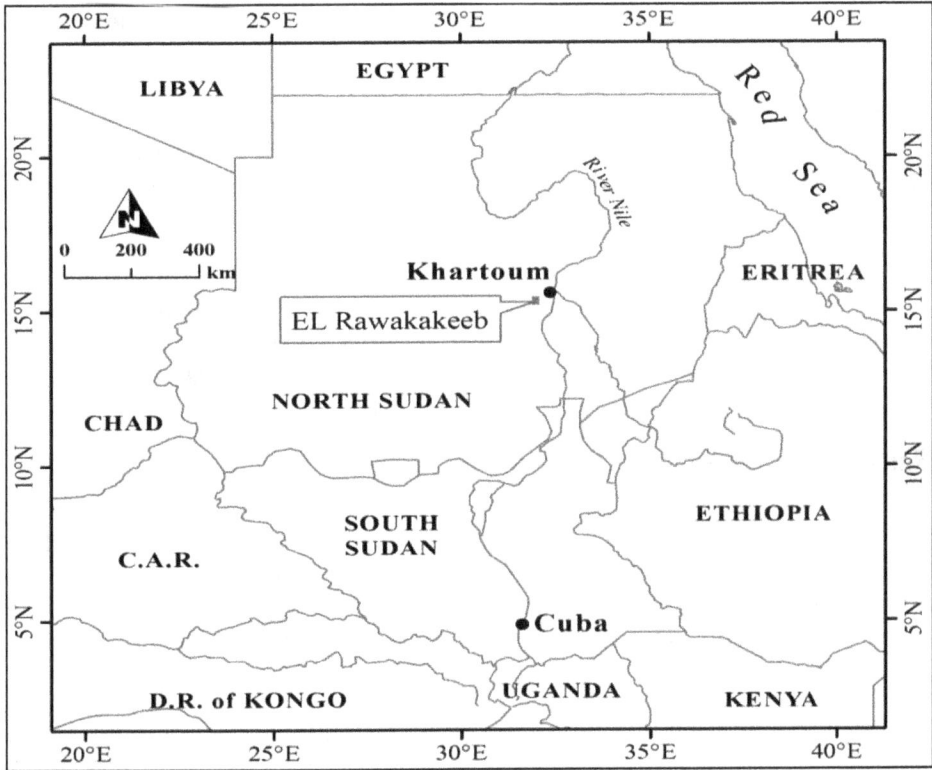

Figure 2. Location Map of RDLRS.

## Soil

ELRawakeeb soil analysis showed that the relative proportions of different soil particles follow the order: sand, silt and clay with sand comprising the highest proportion. Chemically, EL-Rawakeeb soil is moderately alkaline, very poor in nitrogen and carbon, moderate in its bicarbonate and potassium contents and rich in its sodium, calcium and chloride contents (ElHag *et al.*, 2006). According to USDA Taxsonomy, (1975),the soil falls in the order of Aridisols, mixed Koalintic isohyper thermic Gypsic or Typiccamborthid.

## Vegetation

Ahmed (1997) reported that because of high degree of temperature, scarcity of rainfall, natural vegetation is scattered, however, Acacia species are dominant beside some annual shrubs and grasses which grow in rainy season. Recently, there are two rows of eucalyptus and *Acacia mellifera* established as shelter belts.

## Sand Dunes

The prevailing windstorms resulted in sand dunes and sand sheets formation that scattered all over the project area. Three distinct sand dunes are recognized along and opposite irrigation ditch having the following dimensions (length and height respectively), L07mx 2.l0m, 66mx3.30m and 15x3.60m (Naash, 2009).

## Land Use System

Land use system was mainly pastoral. Traditional agricultural activities were usually practiced in low situated areas. Since early seventies, both activities were abandoned due to drought and desertification (Ayers and Westcot, 1995).

## Water Resources

The local community depends on undergroundwater and rainfall water for domestic use. There were three bore holes in the project area to pump water; namely North, East, South East and West bore holes (Agabna *et al.*, 2002), but now only the South East bore hole is on work, the others, are no longer used (Naash, 2009).

# Research Projects at DRI

Reviewing the environmental situation at El Rawakeeb, a scientific committee formulated by the National Centre for Research adopted a research approach to fill- in the gap of research in Sudan efforts to control desertification. This approach consisted of two steps: the first is to collect baseline data on the research area, and the second is to conduct applied research on methods of reclamation of desertified areas.

## Baseline Data

Baseline data collected included the following data sets:

★ Climate (mainly rainfall)

★ Soils and geomorphology

★ Biodiversity (fauna and flora)

★ Water resources

★ Socioeconomic conditions

**Applied research conducted:** Research themes included the following:

★ Testing the different methods of afforestation including establishing shelter belts, community forests and introducing Agroforestry practices.

★ Water discharge and appropriate irrigation technologies that could be applied in desertified areas.

★ Soil fertility restoration by application of biofertilizers and organic compost.

★ Breeding for early maturing and drought resistant crops.

★ Supplementary feed for livestock production

★ Biodiversity assessment and conservation.

★ Enrollment of local community in desertification control.

While some elements of the methodlogy were provided, the paper fails to clearly state the steps that were followed in execution of the study and subsequently answering the objectives of the study.

# Results

## Baseline data

Generally, the baseline data revealed that rainfall is below the long term average by almost 30 per cent as a result of the repeated droughts. Soils are poor in nitrogen and carbon contents. Sand sheets and multi-directional dunes dominate the area. The potential undergroundwater, which could be discharged using available water pumps was just enough to irrigate about 30 acres. Vegetation cover is mainly composed of Acacia species and some annual grasses that grow in the lower lands during the short rainy season (Table 13.1). The table also indicated poor biodiversity and low nutritive value of range flora at El Rawakeeb area.

Table 13.1: Baseline Research Projects Conducted at El Rawakeeb and their Results during the Period 1992-to-Date

| Project Title | Results |
|---|---|
| Characterization of El Rawakeeb climate | Short rainy season (2-3 months.), |
| | ✰ Average rainfall ranges between (100-200 mm) and estimated to decrease by 30 per cent |
| ✰ Rainfall characteristics and their relation to Nino phenomenon and water management | The pattern of change in ground water level followed plant water uptake and showed two peaks: before sunrise and around midday |
| ✰ Providing satellite images database on seasonal valleys discharge and rain fall | |
| Assessment of water for irrigation | Ground water: |
| | ✰ Level is 80 m deep, |
| | Slightly saline |
| | ✰ Sufficient to irrigate 30 Fedans. |
| Classification an characterization of El Rawkeeb soil | ✰ El Rawakeeb soil is classified as sandy clay loam, poor in nitrogen and carbon contents, |
| | ✰ Non saline, non sodic soil with low contents of nitrogen and phosphorous. |
| | ✰ Micronutrients level is neither deficient nor toxic |
| | ✰ Moderate hydraulic conductivity and high infiltration value |
| Assessing diversity of El Rawakeeb flora | ✰ Mainly scattered plants of Acacia species, (*e.g.* Acacia tortilis) |
| Studying diversity of El Rawakeeb fauna | ✰ Identification of 3 orders and 8 families of soil fauna |
| | ✰ Identification of 2 species of desert rodents, |
| | ✰ Identification of 13 orders and 18 families of arthropod fauna |
| Identification of diversity of soil microorganisms | Identification of bacteria that utilize organic carbon and nitrogen, |
| Determination of the nutritive value of El Rawakeeb range flora | Crude fiber content ranged from 8.21 per cent - 44.9 per cent indicating the low potentiality of these flora as fodder, |

## Applied Research Data

Results of applied research activities conducted at El Rawakeeb showed that application of biofertlizers to some leguminous and fodder crops to substitute for low nitrogen and carbon contents increased productivity by 40 – 70 per cent. Breeding pearl millet – main food in the project area- for early maturing and drought resistance resulted in a drought tolerant genotype.

As water seepage in sandy soil is usually high; irrigation canals at El Rawakeeb were lined by a cementing material that increased irrigation efficiency by 40 per cent. The whole cultivated area was protected by a shelter belt mainly from indigenous plants and some legumes *e.g.* pigeon peas Cajanus cajan. Other important applied research results are summarized in Table 13.2. Currently, the research is focused on change detection which has occurred as a result of human intervention. Among the changes detected are groundwater level as a result of invasion of prosopis species, increasing soil salinity, increasing biomass productivity, increasing diversity of decomposers and improving socioeconomic status. Improved livelihood is noticed in terms of provision of water for domestic use, which assists population to save time spent seeking for water. It also, helps to provide fodder, gas cylinders and solar electrification, which empower education and health services. Other benefits include improved socialization in the rural community. Bottom up approach to combat desertification at El Rawakeeb was followed by increasing the participatory role of the local community, as shown in Figure 13.3.

### Table 13.2: Applied Research Projects Conducted at El Rawakeeb Area during the Period 1992-to-Date

| Project Title | Activities | Results |
|---|---|---|
| Soil reclamation | ★ Application of biofertilizers | Improved soil properties |
| | ★ Addition of rock phosphate | Increase in plant production by 40-90 per cent |
| | ★ Addition of organic matter (compost) | Tillage treatment increased soil bulk density and moisture content and decreased soil aggregate stability |
| | ★ Study the effect of tillage on three soil physical properties | |
| Plant breeding | Evaluation of impacts of water stress on productivity of Peral millet (*Pennisetum glaucum* L.) | ★ Significant variations in pearl millet (Pennisetum glaucum L.) grain yield and yield component regarding randomly selected millet genotypes |
| | | ★ Plant breeding of a genotype that can tolerate drought |
| Stabilization of sand dunes | Construction of shelter belts | Reduction of sanding and invasion of sand dunes in the project area |
| Improving irrigation efficiency | Lining irrigation canals using local materials | A reduction of 30 per cent of water loss by lining the irrigation canals and conservation of 40-50 per cent of lining fees by using local materials |
| | Detecting change in ground water level | Ground water level is lowered by invasive plants (*e.g. Mesquite, Prosopis juliflora*) |

| Project Title | Activities | Results |
|---|---|---|
| Animal production | Evaluation of the impact of water restriction on desert animal performance | Water restriction induced a significant effect on the apparent digestibility coefficients of dry matter, organic matter, crude protein, crude fiber, nitrogen free extract and total digestible nutrients in Nubian Goats and Sudan desert sheep. |
| Biodiversity | Investigation of the pattern of the spatial variations of soil seed bank as indicated by topography | Depressed soil is richer in seed density than transitional and hill soils. |
| | | Biomass productivity and carrying capacity varies with latitudes and tend to increase in low lands. |
| | | Desert plant litter varies in species relative density with topography and is generally tend to be low. |
| | Identification of rare desert plants and factors that hinder their germination | Seed hardness is noticed to be the key factor that hinders plant germination. |
| | | Treating seed with a heated needle and scratching them with a knife are best methods, |
| | Assessment of decomposers fauna as affected by application of agricultural practices. | Decomposers fauna are positively affected by applictaion of farm yard manure and negatively by the addition of neem leaf powder as biocide |
| Range management | Detecting annual change of range resources and fodder quantities at El Rawakeeb dry land | 13.5 k/hec Fodder was found and low association between this quantity and plant height and root system of the Acacia trees |
| Socioeconomic studies | Assessment of socioeconomic impact of establishing El -Rawakeeb project | Improved food secuirty and living standard |

They take certain tasks to sustain family life, such as provision of water for domestic use, collection of fuel wood for cooking. Some of these activities have an adverse environmental impact and could intensify the desertification process. Women are the ones who face desertification and suffer directly and thus are the ones who accept the idea and can participate directly.

Awareness seminars were done through direct communication with women by gathering them and tell them about the causes of desertification, as well as the methods needed to mitigate the desertification process (Figure 13.4). Women are trained to restore plant cover by planting seeds suitable to be grown in a desert environment. They are also trained to prepare seedlings in the nursery, and transfer them after hardening, to be planted as shelter belts. Gas cylinders with cookers are distributed as the alternative fuel source. Women enrollment showed positive effects on desertification control where plant cover was restored and consequently wind speed and sand creep decreased.

**Figure 13.3: Community Participation in Desertification Control at RDLRS.**

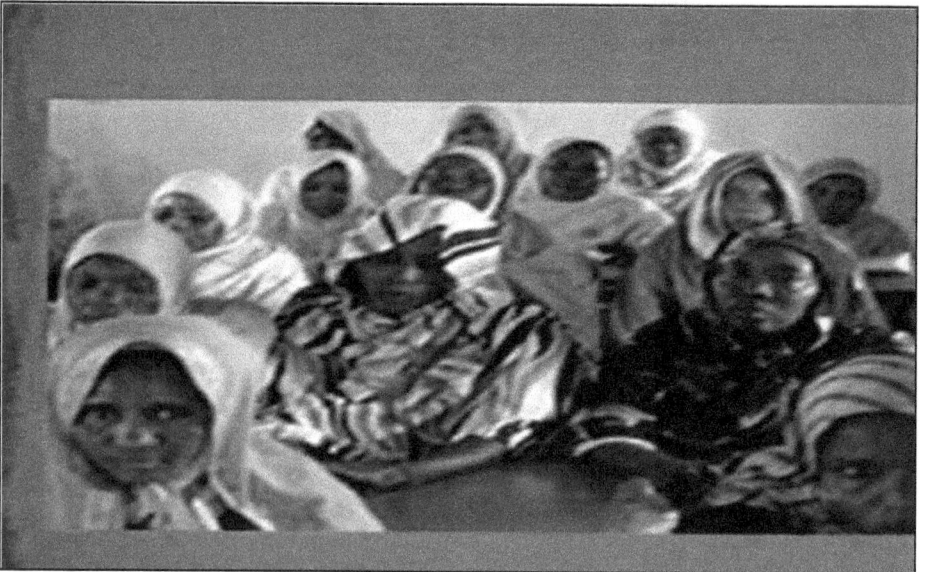

**Figure 13.4: Informal Seminars at RDLRS.**

Relations and co-operations this can be presented as institutional arrangements and stakeholder involvement in desertification control and then you mention the roles, responsibilities and acheivements at different levels

## 1. At Local Level

★ Universities (*e.g.* Khartoum, Sudan for Science and Technology, Shendi, Bakhat el Ruda.etc).

★ Ministries (*e.g.* Agriculture, irrigation, Environment, forestry and physical development,)

★ Organizations (*e.g.* forestry, meteorology Sudan Water Partnership *etc*)

## 2. At Regional Level

★ IGAD, ALESCO, FASRC,ACSAD...*etc.*)

## 3. At International Level

★ UNDP. FAO, TWAS, ARLRC

## DRI Community Services

### 1. Services to Local Community

★ Awareness about desertification issues

★ Provision renewable energy sources (solar energy and biogas)

★ Community participation in reforestation

### 2. Services to Scientific Community

★ Train personnel and groups

★ Act as a platform for researchers

★ Provide modeling site applicable to similar ecological areas

## RDLRS Constraints

Include soil salinity, deep water table, erodible soil surface, wind erosion and man misuse of resources.

## Financial Constraints

Such as lack of logistics for field work and qualified technical personnel.

## Recommendations

★ Upgrade the indigenous knowledge to combat desertification.

★ Direct development programs for protecting genetic resources and wildlife.

★ Develop a model applicable in similar environments

★ Improve the infrastructure to include laboratories, rest house, conference rooms.

★ Strengthen cooperation at local as well as international levels

★ Provide a scientific platform for scientists

★ Secure funds for future expansion

## Discussion

Sudan is one of the most seriously affected countries by desertification in Africa. The arid and semiarid lands cover an area representing about more than 90 per cent of the total area of the country.More than half of its population lives in these areas. The recent increase of population has led to the decline of the natural resources of the semi-arid zone of the Sudan, and consequently economic status has been affected. Sudan has attempted to combat desertification and mitigate its effects, but these efforts failed to achieve the planned objectives because of several factors among which is lack of scientific research. Recently, after the success achieved at El Rawakeeb, other academic institutions have established similar centers for research in desertification control *e.g.* University of Khartoum, Gezira University.

Research activities at El Rawakeeb area succeeded in clarifying some of the desertification causes, including environmental factors and human actions. Environmental factors have led to changes in soil properties manifested in the formation of sand dunes and sand sheets. These results coincide with those of Laki,(1994), Elhag, (2006) and Khairalseed, (2015), who studied desertification causes and effects on different parts of Sudan. Soil reclamation efforts improved soil characteristics such as increased soil fertility by 25-40 per cent and crop productivity by 45-75 per cent over time. Repeated monitoring of vegetation in the area revealed that almost all types of vegetation cover exhibit fluctuating patterns. However, afforestation and Agroforestry studies succeeded to restore plant cover by 60 per cent. Studies in water management showed that degradation of the lands was invoked by poor techniques of water harvesting and spreading. Improving irrigation technologies increased irrigation efficiency by 60 up to 90 per cent.

As plants are crucial for soil and water conservation and biodiversity protection, certain areas were planted to fix sand dunes. As part of desertification control,local community was connected to this initiative, with aim to restore plant cover. Plant breeding is of paramount importance in arid and semi arid regions if expected climate change is considered. The results obtained at ElRawakeeb research station encourage breeding for drought resistent and/or quick maturing crops in all sub Saharan African countries. This may include breeding palatable species of vegetation for millions of animals living in dry lands specially if we consider the general pattern towards drier and warmer climates since 1970s. Abdi *et al.* (2013).

As animal production is an important subsidiary activity for rural inhabitants, research on animal performance in desertified regions is crucial. Sudan desert sheep and Nubian goats were bred under heat and water stresses at El Rawkeeb station. These animals showed positive responses to stress. Such responses are seen in terms of increased dry matter intake, water intake, fecal dry matter and respiratory rate.

As it is cited in Ayoub, 1998, wind erosion was the most widespread soil degradation type in the arid zone. Rizgalla, (2013) illustrated that the removal of plant cover from sandy soil surfaces at El Rawakeeb area has exposed the soil surfaces to significant losses by wind erosion.

As mentioned above (see 2.1.), desertification induced poverty and food insecurity, (Suliman and Salih, 2006; Mohamed and Salih 2007). In addition to that, it manifested in increased poverty, illiteracy, spread of some diseases, increased rate of infant mortality and morbidity. Lack of water for domestic use caused a heavy burden on families (Elias *et al.,* 2008). Water provided by the station to the local community has uplifted this burden.

Since women are the primary custodians of the indigenous knowledge system, awareness seminars to empower women and increase their role in combating desertification were organized. As a result, they participated significantly in restoring plant cover and preventing tree cutting by using alternative energy,(Gas cookers), (Abdel Latif and Diab, 2013).

## Conclusion and Recommendations

Desertification is a serious problem confronting Sudan. It results from interaction of natural and human factors. It is intensified due to lack of scientific oriented methods to combat it. El Rawakeeb site was chosen by the National Centre for Research to be a model site for desertification control. The ecological features of this site contributed positively to such a choice. An interdisciplinary research team working at the National Centre for Research was assigned to take responsibility of conducting research on desertification control. This team started with collection of database on climate,Hydrogeology, soils, flora and fauna in addition to socioeconomic parameters. These databases were used to formulate applied research in desertification control and mitigation of its consequences. Some of these research projects included: water management,soil reclamation, sand dune fixation, breeding for drought tolerant plants, and animal husbandry.

Based on the scientific approach and the encouraging research results, it is recommended to:

★ Consolidate research in Sudan to cover area suffering desertification.

★ Transfer the relevant research results to the local community at El Rawakeeb with the aim to improve their socioeconomic situation.

★ Consider El Rawakeeb research station as an educational model for researchers and students interested in desertification control.

★ Some nice findings and discussions captured in the body of the paper have not been provided as recommendations for example; researcwh on effects of environmental factors on soil properties, the effects of plants on soil and water conservation and biodiversity protection.

## Acknowledgement

The authors are gratefull to the National Centre for Research for funding and supporting the conducted research project. They also, apprecitiated the help efforts from the Documentation Centre (NCR), and librarian at NCR and related research institutes.

# References

1. Abdel Latif M.A. and Diab, E. E. (2013). Empowering women skills in combating desertification in Sudan. Proceedings of the international conference on empowering women in developing countries through information and communication technologies, organized by Jaypee University of Information and Technology and Centre for Science and Technology of the Non- Aligned and other developing countries (NAM S&T Centre), New Delhi, India, 1-3June 2013.

2. Abdi, A. Glover, E. and Luukkanen, O. (2013). Causes and impacts of land degradation and desertification: Case Study of the Sudan. International Journal of Agriculture and Forestry, 3 (2): 40-51.

3. Agabna, E. and Abdel Rahman, H. (2002). Quality of boreholes water in El Rawakeeb dry land, Desertification Research Institute, National Centre for Research, Annual Report, Khartoum, Sudan.

4. Ahmed, I.M. (1997). The Present flora of El-Rawakeeb area. M.sc. Thesis, faculty of science, University of Khartoum

5. Ayers, R. S. and D. W Westcot. (1995). Water Quality for Agriculture Organization (FAO) of The United Nations. FAO Irrigation and Drainage Paper no.29 Rev.1

6. Ayoub A.(1998).Extent, severity and causative factors of land degradation in the Sudan.Journal of arid environment, 38(3): 397- 409.

7. Bauer, S. (2007). "Desertification". In Thai, Khi V. *et al.*, Handbook of globalization and the environment. CRC Press. ISBN 978-1-57444-553

8. Briassoulis, H. (2005). Policy integration for complex environmental problems: the example of Mediterranean desertification. Ashgate Publishing. p. 161. ISBN 978-0-7546-4243-5.

9. Cornet A., 2002. Desertification and its relationship to the environment and development: a problem that affects us all. In: Ministère des Affaires étrangères/ adpf, Johannesburg. World Summit on Sustainable Development. 2002. What is at stake? The contribution of scientists to the debate: 91-125.

10. Dobie, P. 2001."Poverty and the drylands," in Global Drylands Imperative, Challenge paper, Undp, Nairobi (Kenya) 16 p.doi: 10.1016/j.jaridenv.2008.06.010.

11. Elhag, M. M. (2006). Causes and Impact of Desertification in the Butana Area of Sudan. Ph.D. thesis, faculty of Natural and Agricultural Sciences, University of the Free State Bloemfontein, South Africa.

12. Elias, O. Eldalel, N. and Elnasikh, M. (2008). Impact of desertification on food security and man poverty at El Rawakeeb dry land. Desertification Research Institute (DRI) annual scientific report published by the National Centre for Research, Khartoum, Sudan, pp. 45-62.

13. Geeson, N. (2002). Mediterranean desertification: a mosaic of processes and responses. John Wiley and Sons. p. 58. ISBN 978-0-470-84448-9.

14. Geist, Helmut (2005). The causes and progression of desertification. Ashgate Publishing. ISBN 978-0-7546-43234 www.http: //books.google.com/ books?id=acbWdynlU3cC.

15. Khairalseed, A. R. (2015). Desertification in Sudan, concept, causes and control. ARPN journal of science and technology, 5 (2): 87 – 91.

16. Laki, S. L. (1994). Desertification in the Sudan: causes, effects and policy options. The international journal of sustainable development and world ecology, 1(3): 198-205.

17. Longjun Ci and Xiaohui Yang, (2010). Desertification and Its Control in China. Springer. p. 10. ISBN 978-7-04-025797-7. http: //books.google.com/ books?id=agd8MFDYLXEC and pg=PA10

18. Mares, Michael S., ed. (1999). "Middle East, deserts of". Encyclopedia of deserts. University of Oklahoma Press. pp. 362. ISBN 978-0-8061-3146-7. http: //books. google.com/books?id=g3CbqZtaF4oC and pg=PA362

19. Millennium Ecosystem Assessment (2005) Desertification Synthesis Report.

20. Mohamed, S. and Salih, A. (2007). Assessing environmental degradation on the socioeconomic status at El Rawakeeb dry land. Desertification

21. Research Institute (DRI) annual scientific report published by the National Centre for Research, Khartoum, Sudan, p 63-76

22. Naash,M.E.(2009). Assessment of land degradation using some biophysical indicators at EL-Rawakeeb development project. M.Sc. Thesis,faculty of agriculture, University of Khartoum

23. Parrillo, V. (2008). "Desertification". Encyclopedia of social problems, Volume 2. SAGE. ISBN 978-1-4129-4165-5 http: //books.google.com/books?id=mRGr_B4Y1CEC and pg=PT271

24. Pasternak, D. and Arnold, S. (2001). Combating desertification with plants. Springer. p. 20. ISBN 978-0-306-46632-8. http: //books.google.com/books?id=B-i8-DPf6xgC and pg=PA20

25. Rizgalla, F.,Elhadi, E. and Abdalla, M. (2013). Soil organic matter size fractionation in desert tropics as influenced by application of organic residues. Ann. Repo. pp. 168-176.

26. Stringer, LC (2008). Review of the international year of deserts and Desertification 2006: What contribution towards combating global desertification and implementing the united nations convention to combat desertification?, j.jaridenv, 72(11), pp. 2065-2074.

27. Suliman, A. and Salih, A.(2006). Assessment of the impact of desertification on food security and poverty at El Rawakeeb dry land. Desertification Research Institute (DRI) annual scientific report published by the National Centre for Research, Khartoum, Sudan, pp. 77- 82.

28. United States Geological Survey, "Desertification", 1997.

29. UUNCCD (1992). United nations convention to combat desertification. In those countries expressing serious drought and/or desertification, particularly in Africa. Secretariat of the UNCCD, Bonn, Germany, Whitford, W. G. (2002). Ecology of desert systems. Academic Press. p. 277. ISBN 978-0-12-747261-4.

30. World Bank (2009). Gender in agriculture sourcebook. World Bank Publications. p. 454. ISBN 978-0-8213-7587-7.

## Chapter 14

# Capacity Building and Criticality of People's Participation in Drought Management and Desertification Control in Uganda, East Africa

### Kisamba Mugerwa Wilberforce

*PhD, Executive Chairman,*
*National Planning Authority, Uganda*
*E-mail: wkisambamugerwa@npa.ug, wkisambamugerwa@gmail.com*

## Abstract

There is growing concern among scientists and development analysts that drought and desertification are at the core of serious challenges and threats facing development and human life in Africa. This concern is based on the associated problems that have far reaching adverse impacts on human health, food security, economic activity, physical infrastructure, natural resources, the environment, and national and global security. Drought and desertification affects a large number of pastoralists who constitute over 20 per cent of people in Uganda's dry lands, in what is known as the 'Cattle Corridor'. It affects the vegetation in rangeland areas which is the principal source of livelihood for their livestock and themselves.

A number of changes have taken place in the management of rangeland resources in Uganda, with subsequent governments and development agencies investing substantial resources in an effort to improve rangeland productivity. However, all these efforts were carried out without any systematic programs forinvolving pastoral farmers in sustainable development of their rangeland in theCattle Corridor. These policies and development interventions reduced the pastoral land area available to traditional pastoralists, and became a source of conflict, both within the cattle corridor and also with neighboring districts.

Against this background, this paper addresses the issue of capacity building and the criticality of people's participation in drought management and desertification control in Uganda with a particular focus on the rangeland areas based on the case study in the 'cattle corridor'. It examines the interventions so far implemented to address the problem of drought management and desertification control in rangelandareas in Uganda. It specifically addresses the criticality of people's participation and efforts towards capacity building to ensure acquisition of skills for other emerging opportunities. Data was obtained through a review of the findings of earlier papersand government documents like the white papers, policy documents and legal records. The findings of this paper show that in executing mitigation and adaptation measures for drought management and desertification control, the participation of people has been neglected. The study recommends that the local communities should be mobilized, educated and skilled, so as to constructively participate in the sustainable management of natural resources in the rangelands.

*Keywords: Drought, Desertification, Participation, Rangeland.*

## Introduction

Drought and desertification remain some of the serious challenges and threats facing sustainable development in Africa, more especially the pastoralists in the rangelands in Africa in general. The increasing severity of desertification and droughts and their recurring nature have eroded the livestock assets ofpastoralists.

The frequent occurrence of droughts and its impact on thepastoral communities are well understood and documented.The central element in drought and desertification is water deficit. Simply put, drought is an extended period such as a season, a year, or several years, of deficient rainfall relative to the statistical multi-year average for a region(Holden and Shiferaw, 2004). Desertification, on the other hand, is regarded as a process of land degradation in arid, semi-arid and dry sub-humid areas, resulting from various factors, including climatic variations and human activities(UN Convention to Combat Desertification, 1994).Desertification is caused in a way that dry lands ecosystems are extremely vulnerable to over-exploitation and inappropriate land use (Hulme and Kelly,2013).

Much argument has been directed to the notion that over-cultivation, inappropriate agricultural practices, overgrazing and deforestation are the major causes of drought and desertification. In essence, however, it is an amalgamation of changing weather conditions which result in a reduction of rainfall and the much deeper underlying forces of socio-economic nature, such as poverty and total dependency on natural resources for survival by the poor(Rowell *et al.*, 2012; Yang and Prince, 2010).

However, despite the technical knowledge and documentation of the impactof desertification and drought on thepastoral communities, focus on the measures to ensure pastoral community preparedness to counter desertification and drought measures hasbeen given little attention.To become effective in dealing with desertification and drought, government and other stakeholders require serious enhancement of local pastoral communities' preparedness,through capacity building and people's participation in sustainable use of local natural resources in the rangelands. The pastoral methods of seasonal mobility andtranshumance dimensions must not be regarded as a negative way of life that must be changed, but it should be considered and incorporated in desertification and droughtrisk reduction and adaptation planning. For this to happen,government and other stakeholdersneed to improve capacity building and people's participation in the implementation of desertification and drought riskreduction activities.Such cooperation and resultant synergies are aimed at buildingthe capacities of local pastoral communities to prepare for and adequately respond toemergencies as well as enhancing livelihood assets ofpastoralists and their respective areas.

In Uganda, drought and desertification are receiving increasing attention in development circles. Acute drought and desertification have most especially affected severely various sections of the people in Uganda, more especially the pastoral people in the rangelands in what is known as the 'Cattle Corridor'. Drought and desertification have brought about severe repercussions to the pastoralists like loss of livestock, land conflicts and loss of rare vegetation species.

However, the neglect of pastoral people's participation in solving the problemsfacing them in the rangelands is seen as the most critical concern in drought control and desertification management. The criticality of people's participation is that as decreasing grazing land cannot continually support the increasing livestock populations, conflicts have ensued between the pastoralists and the people pastoralists presume to have encroached on their grazing lands. Yet, there is little concern from Government to involve pastoral people in its efforts to resolve these matters. Depletion of pastures, increasing social tension and pastoral land use conflicts, and a declining grazing land area have persisted aspoints of unease for the pastoralists, government and the country at large.

Most crucially, land management practices in the rangelands fail at recognizing pastoral people's indigenous knowledge and their participation in resolving their own problems. The objective of this paper therefore is to examine the issue of capacity building and the criticality of people's participation in drought management and desertification control in Uganda with a particular focus on the rangeland areas based on the case study in the 'cattle corridor'.

## Methodology

The findings of this paper were obtained through a review of empirical evidence from various papers on the subject in Uganda. Government documents like the white papers, policy documents and legal records were also reviewed and analyzed for findings for this paper. Most of the findings were derived from an earlier study on "Rangeland tenure and resource management: An overview of pastoralism in Uganda" by Kisamba Mugerwa (1992). The findings of this paper were further scrutinized through content analysis as well as rigorous review of the facts and evidence in earlier papers, and were triangulated with current evident from recently obtained data. A scheme for dealing with drought management and control of desertification is described in Figure 14.1

## Literature Review

### A Brief Overview of Drought and Desertification in Africa

Africa is especially susceptible to land degradation and bears the greatest impact of drought and desertification. Two thirds of Africa is classified as deserts or dry

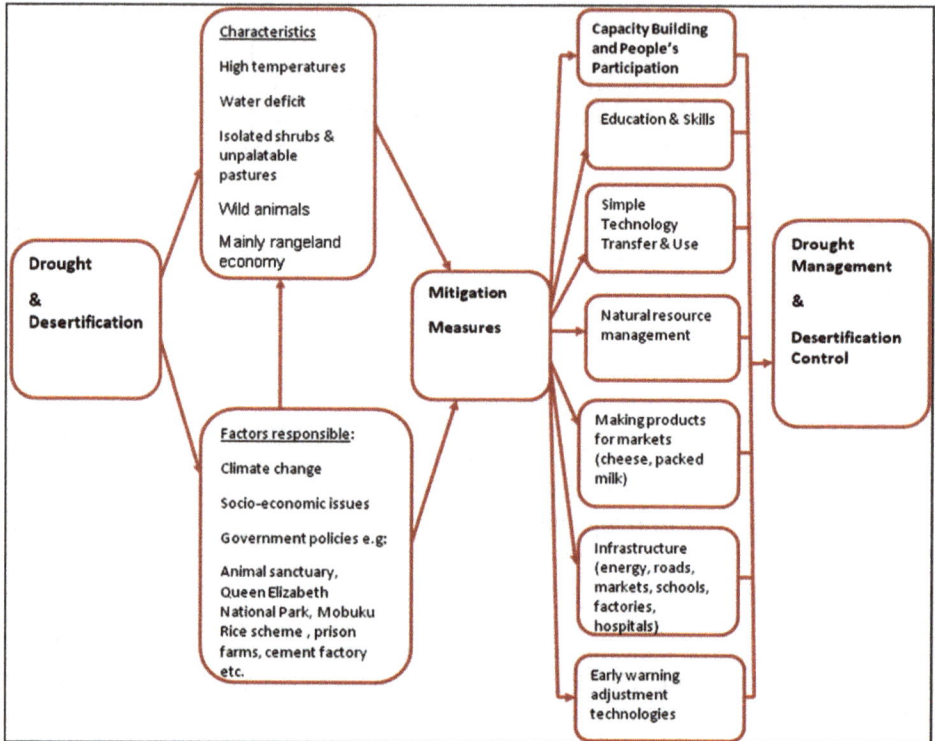

**Figure 14.1: Framework for Addressing Drought Management and Desertification Control in Pastoral Areas in Uganda.**

lands and these are concentrated in the Sahelian region, the Horn of Africa and the Kalahari in the south. It is estimated that land degradation affects at least 485 million people or sixty-five per cent of the entire African population (Riebsame *et al.*, 2005).

Desertification around the Sahara has featured out as one of the potent symbols of the global environment crisis. Climate change is set to increase the area susceptible to drought and desertification in the region. It exacerbates the occurrence of climate related disasters, as current climate scenarios predict that the driest regions of the world will become even drier in many parts of Africa. It is projected that there will be an increase of 5-8 per cent of arid and semi Arid lands in Africa.

The seriousness of drought and desertification in sub Saharan Africa is that rain-fed agriculture provides about 90 per cent of the region's food and feed (Rosegrant, *et al.*, 2002). It is the principal source of livelihood for more than 70 per cent of the population, as over 270 million people in Sub-Saharan Africa live in absolute poverty (Hellmuth *et al.*, 2007). Because of heavy dependence on rain-fed agriculture, about 60 per cent of Sub-Saharan Africa is vulnerable to frequent and severe droughts (Esikuri, 2005).

Estimates from individual countries report increasing areas affected by or prone to desertification. It is estimated that 35 per cent of the land area (about 83,489 km² or 49 out of the 138 districts) of Ghana is prone to desertification (Hellmuth *et al.*, 2007). Seventy per cent of Ethiopia is reported to be prone to desertification, while in Kenya, around 80 per cent of the land area is facing desertification (Gautam, 2006). Uganda, located in East Africa, is one of the countries that are most affected by desertification in Africa. It occupies an area of 241,000 km$^2$ of which 80,000 km$^2$ is dry lands. It has a human population of 35 millions of which 90 per cent live in rural areas (NPA, 2017). Uganda's dry lands, which receive between 500 mm and 1000 mm of rainfall, occupy an area stretching from the north-east through central to the south-west of the country, an area commonly referred to as the "Cattle Corridor"(KisambaMugerwa, 1992). Uganda has been experiencing climate changes, with rising temperatures, which have caused the disappearance of water bodies, and decline in food production since 1990s (see Figure 14.3). According to FAO (2008), over 80 per cent of Uganda will become a desert in less than 100 years if the current climate changes are not addressed.

Over 40 per cent of the pastoralists, who constitute the majority in the country's dry lands, live below the poverty line. Their high population growth rate increases pressure on limited and fragile land resources. This scenario has intensified land and environmental degradation leading to loss of biological productivity of the land resulting in frequent famine, low household incomes, and increased social unrest in the affected areas.Soil compaction, erosion and the emergence of low-value grass species have subdued the land's productive capacity, leading to drought and desertification (Rosegrant, *et al.*, 2002). The"Cattle Corridor", the rangelands that stretch from Uganda's border with Tanzania in the south to Karamoja region in the north border with Sudan and Kenya, is themost affected area by drought and desertification (Figure 14.2).

## Capacity Building and Participation amongst Pastoral Communities in Uganda

In recent years, there has been some clear acceptance of the inevitability of pastoralism as an alternative economy. Rather than settling pastoralists as has been common in the past, education and capacity building in general programs that are able to mobilize the technology and economic efficiency of traditional pastoralists and herding systems are promoted. It is hoped that such approach will improve both pastoralists' livelihoods and potential economic viability by disseminating the advantages of modern methods of agriculture and livestock management.

Policies and programs concerning capacity building through the education of pastoralists, transfer of technology and skills can be grouped around two major rationales which may work together or against each other: (*a*) the full accomplishment of the individual as a human being, and (*b*) the integration of nomadic groups into the wider national context. The first group is centered on a notion of education as a basic need and a fundamental right and puts great emphasis on inclusion and empowerment. The second focuses on the economic and social development of pastoralists. In this case the main concern is with sedentarisation, modernization, poverty alleviation, resource management and state building. In the large majority of cases, this involves the incorporation of pastoralists into mainstream society and

**Figure 14.2: Sun Scorched Vegetation in Karamoja Rangelands in Northeastern Uganda.**

**Figure 14.3: Drought Struck Maize Garden in Uganda.**

economy. Basic formal education is seen as essential for the full accomplishment of individuals as human beings, their survival and lifelong development. This position is reaffirmed for example in the first article of the World Declaration on Education for All (1990). As such, education is represented as a fundamental human right.

Although, the above view offers many advantages, in the specific context of education provision to pastoralists it also presents some dangers. The first is that by focusing on individuals, it separates children's livelihoods and best interests from those of their households, therefore antagonizing the structural organization of the pastoral economy, the basic unit of which is the household or a group of households, not the individual. This clash between individual-based education strategies and household-based pastoral strategies is a crucial issue.

## Empowerment and Inclusion

The satisfaction of basic learning needs is thought to have, as a consequence, the empowerment of individuals. The nature and limits of empowerment within development practices, including education, are the object of critical analysis which are beyond the scope of this study (Shore and Wright, 1997; Nelson and Wright, 1995; de Koning, 1995). With specific reference to pastoralists, the goal of empowerment, through education or anything else, seems particularly appropriate, given that almost in every country where they are found, they are minorities suffering problems of under-representation, social, economic and geographic marginalisation, and incorporation by hegemonic groups. Particularly within mainstream policies, the notion of empowerment is often adopted in conjunction with a model of basic education that assumes development, individual liberty and social mobility as inherently implied by it (Nigeria Federal Ministry of Education, 1987).

## Development and Integration

With rare exceptions, education is seen as an instrument for transforming pastoralists into (i) settled farmers or waged labourers, (ii) "modern" livestock producers, and/or (iii) loyal citizens (de Koning, 1995). Only small scale non-formal education is sometimes used as part of a wider process directed to articulating the internal dynamics of nomadic society with national and global dynamics. Common to all the education-for-development approaches are accounts of pastoralists' poverty and the assumption that education will bring an improvement of their standard of living.

For about sixty years, from the first experiments in the 1920s to the mid-1980s, at the core of pastoral development theories and practices was the assumption that pastoralism was an evolutionary way of life, environmentally destructive, economically irrational, and culturally backward (Anderson, 1999). The only way pastoralists could develop was by stopping being pastoralists, the obvious first step onto a higher stage of evolution being sedentarisation. Indeed, even after more than two decades of extensive research accumulating evidence against such tradition from different disciplines (Sandford, 1983; Scoones, 1995; de Bruijn and van Dijk, 1995; Pratt *et al.*, 1997). Such myths, long been dismissed within academic research, are proving hard to kill off amongst pastoral development policy makers and operators. It is not surprising therefore, to find those myths again and again within some development circles, which is for the most part produced by educationalists, usually without specialist knowledge of pastoralism. Of all the myths, that the development of pastoralism is sedentarisation seems to be the most resistant.

Pastoralists are formally categorized according to their stage of sedentarisation in Nigeria's *Blueprint on Nomadic Education* (Federal Ministry of Education, 1987), and even in a recent study from a participatory perspective one reads: 'Rural sedentarism in Africa can be discerned as the last stage of the process that occurs over time in the mode of the pastoral production. The three stages towards sedentarisation are nomadic-pastoralism, agro-pastoralism and transhumant-pastoralism' (Woldemichael, 1995).

This approach tends to define pastoralism in purely negative terms, with reference to what they are not-yet or not-anymore. They are identified as 'farmers of livestock' (Ezeomah, 1997), represented as virtually settled people but without a place to stay: 'lack of a home of his own and grazing land for his cattle has forced him to be on the move throughout his life' (Alkali, 1991: 56).

There are even debates on the way sedentarisation should occur, whether the process should be accelerated through education provision or rather left to follow its "natural course", for example the papers of G.V. Ardo, J. Aminu, and H. Alkali, in the collection of studies on *Nomadic Education in Nigeria* edited by GidadoTahir (1991), or the critique by Woldemichael (1995) of government-initiated sedentarisation in Eritrea.

However, historically pastoralism is a specialisation that developed from agriculture (for example, Kazhanov, 1984; Sadr, 1991). So there is no evolutionary straight line from nomadic livestock keeping to sedentary farming. The problems associated with the sedentarisation of pastoralists have been the object of detailed analysis over the past twenty years (Salzman, 1980).

## Capacity Building for Modernisation

Following the shift in the pastoral development paradigm during the 1980s (Baxter, 1985; Hogg, 1988; Anderson, 1999), some countries abandoned, at least formally, the goal of sedentarisation and transformation of pastoralists into farmers, beginning to focus on how to use education in order to improve pastoralism as such. Pastoralists should receive formal education because, within their respective countries, they control important "national" resources (land and livestock), the productivity of which should be improved to match national requirements. Education is seen as an instrument to change pastoralists' attitudes and beliefs, as well as to introduce "modern" knowledge and "better" methods and practices. In short,it is to transform nomadic pastoralists into modern livestock producers.

In Ethiopia, for example formal education is supposed to introduce agents of change within pastoral communities. Usually a concern for environmental degradation, deterioration of pasture or desertification is associated with a concern for productivity levels. As a result, increased marketing and marketability of livestock are included in education policies and even curricula.

In Tanzania, the Ministry of Education and Culture emphasises the urgency of educating pastoralists on the need to decrease the size of their herds in order to reduce the pressure on the land. The argument goes on to recommend the application of modern methods of animal husbandry, such as the use of better cattle feeds, preparation of fodder and pasture management, with the goal of improving animal products for wider markets (Bugeke, 1997: 78).

Despite the shift away from the emphasis on sedentarisation, education continues to be intended as an instrument for the transformation of pastoral society, although this time 'from within', in order to modernise pastoralists 'without uprooting their culture' (Ezeomah, 1983). However, the attention given to indigenous culture reduces it in practice to a stock of 'essential elements' identified with the help of consultations with the pastoralists but ultimately chosen by experts (educationalists), to be blended or incorporated into the 'nomadic education' curriculum with the explicit intent of making schooling more appealing to the pastoralists (Salia-Bao, 1982; Lar, 1991).

The many claims about the beneficial effects of education on pastoral productivity are not supported by evidence. Indeed, very little research has been carried out on the subject and the few data available appear to disprove the argument rather than support it. On the other hand, among the Maasai of Kenya, who have increasingly turned to school education during the last twenty years (Sarone, 1986), education does not affect livestock production, which is being taken over by young non-educated wealthy cattle traders who buy the labour of young non-educated stockless herders (Holland, 1996).

Moreover, a crucial assumption in all the approaches emphasising the role of education in increasing productivity is that it is possible to separate pastoralism as a way of life from pastoralism as a way of production, abandoning the first in order to modernise the second. This perspective assumes that the individuals will be "emancipated" through education from their traditional way of life as pastoralists, but will maintain the same productive role as herders. However, in practice this hasn't proved to be the case.

The reduction of pastoralism to a system of production didn't work when it was first theorised and experimented by the Soviet Committee of the North in Siberia in the 1930s, on the Tungus (Evenki) reindeer nomadic pastoralists. As a consequence of forced collectivisation of livestock and division of labour, only the people in hunting and herding brigades were to work away from the settlements, with one woman each as housekeeper. But rather than staying in the settlements many women followed their husbands taking with them those children who didn't have to stay in boarding schools (Habeck, 1997).

Anthropologists have pointed out how pastoralism is a mode of perception as well as a mode of production (Baxter, 1990). An awareness of the non-viability of pastoral livelihood strategies in the face of shrinking resources and lowering social status appears to trigger in pastoralists a concern for their own existence and cultural identity rather than an economic concern about the necessity of modernising their production methods. The responses sought to such a concern may have more to do with spirituality than economics.

## Findings and Results

### Pastoralism in the Rangeland in Uganda

Pastoral or range lands in Uganda vary in their physical, biological, and climatic nature as well as human activity dimensions. But as a useful generalization,rangelands are wild grasslands in which grasses are originally not artificially seeded and are managed as permanent pastures. They are still characterized by high temperatures, low and highly variable rainfall regimes, low vegetation cover density and fragile soil. The main economic activity is pastoralism. However, pastoralism and wildlife in form of game reserves and national parks mainlygo hand in hand for the tourism industry.

The rangeland areas cover the 'Cattle Corridor' stretching from Uganda's border with Tanzania to the Kotido and Moroto Districts in northeast Uganda (see Figure 14.4). The area covers parts of Ibanda, Mbarara, Kiruhura, Rakai, Masaka, Lwengo, Sembabule, Bukomansimbi, Kalungu, and Kasese Districts, Kyaka County of Kabarole District, parts of Kibale and Mubende Districts, Ngoma Sub-county and Nakasongola District, Ngoma District, Luwero District, Baale County in the northern part of Mukono District, the eastern parts of Masindi District, the northern parts of Kamuli District, and finally extends through parts of Apac, Lira, Soroti, Katakwi, Kotido, Amuria, Nakapiripirit and Moroto districts (KisambaMugerwa, 1992).

These areas are mostly semi-arid or arid. The main cattle keepers are the

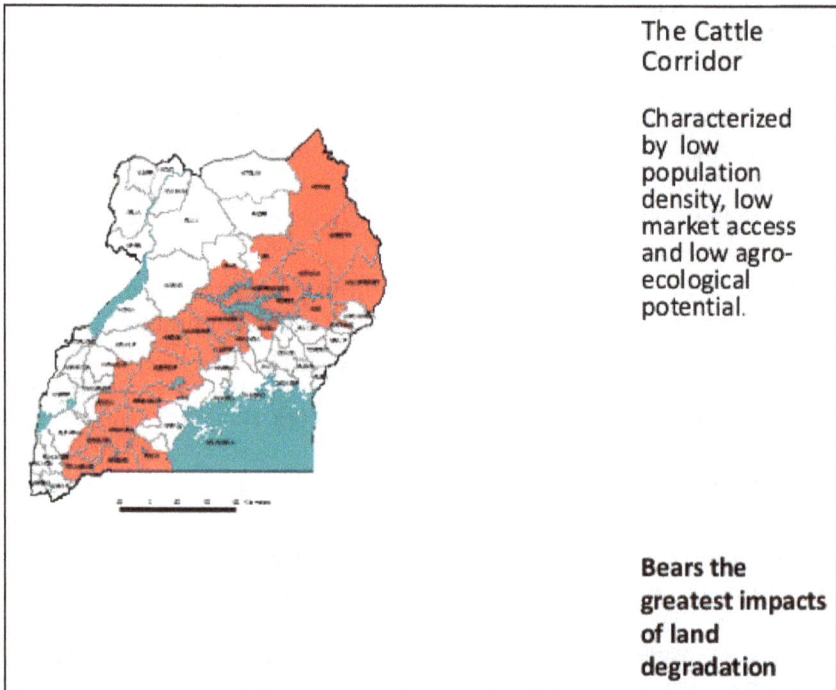

Figure 14.4: Map Showing Cattle Corridor in Uganda.

Bahimatribe in Ankole area, the Basongoraon the foots ofRwenzori Mountains in Kasese District, and the Karimojong in the northeast. Other cattle keepers in the area include the Itesot of Soroti District, the Baruli of Nakasongola County in Nakasongola District, and those of mixed ethnic background in Mubende, Luwero, Masaka, and Masindi Districts. They have similar political economies, though the Itesots and the Baruli practice sedentary cattle keeping, and the Basongora and Karamojong practice transhumance (KisambaMugerwa, 1992).

The pastoral/range areas generally experience bimodal and unreliable rainfall with a long dry spell from October to March. The mean annual rainfall varies from approximately 500 mm to approximately 1,000 mm with high levels of fluctuation between years and among sites. The temperature averages between 18°C and 20°C with a maximum of 28°C to 30°C (UBOS, 2016). Warm temperatures and unreliable rainfall combine with scorching winds during dry spells. In some areas, particularly the southwest part of the corridor, the climate tends to improve and rainfall may reach as much as 1,125 mm a year. Temperatures vary daily from 2°C to 7°C. This type of climate is usual in wooded savannahs (UBOS, 2016).

Cattle provide a major source of food and a means of livelihoodin the 'Cattle Corridor'. Traditionally in Karamoja, cattle herds are built through raids. The Zebu type of cattle, a local breed with short horns, dominates. Recently, cattle raids have become an integral feature of life in Karamoja with cattle passing among different groups and clans, providing the ascendant group(s) of the moment with a means of livelihood and resulting in the marginalization of the defeated group(s). The widespread acquisition of automatic weapons by Karimojong raiders had increased the violent nature of cattle rustling, creating deadly cattle raiding among neighboring tribes and among different ethnic groupings.

With some ethnic groups being marginalized, the Karimojong economy is increasingly characterized by semi-nomadic or transhumance pastoralism, supplemented with some agricultural mono-cropping. In the southern ·and western parts of the Karamoja region that border other districts in Uganda, the population is more sedentary, and agriculture plays a major role in the economy. In this paper, it is estimated to be more than 50 per cent of the population in Karamoja is settled, and about 30 per cent depend more on cultivation than livestock.

The shift from livestock is partially attributed to the inter-ethnic raids, and 50 to 60 per cent of the households in Karamoja do not have any cattle. Among the ethnic groups that have abandoned cattle keeping are the Pokoth, Chekwe, Labwor, and Dodoth. The Jie, Matheniko, Bokora, and the Pian still depend heavily on a pastoral economy. In 2015, estimates listed the number of cattle at about seven million. Goats and sheep are generally not raided and outnumber cattle.

In southwest Uganda, the Bahima live in an area formerly known as Ankole and which now covers the districts of Mbarara, Ibanda, Kiruhura, Bushenyi, and Ntungamo. They traditionally herd the long-homed Ankole cattle which account for about 20 per cent of the cattle in Uganda (KisambaMugerwa, 1992). Milk is the main product and the number of cows is deliberately kept high to ensure a consistent supply (KisambaMugerwa, 1992). Pasture land is traditionally the communal property of the tribe, and members of the tribe can graze as many animals as they like. They used to move to other areas for better grazing land and to search for water for the animals.

However, due to increasing difficulties caused by solo dependence on livestock by some pastoral communities in Uganda, especially the Karimojong, the pastoralistsare taking to cultivation (KisambaMugerwa, 1992). Open cultivated parcels of land around the towns of Kotido and Moroto have become a permanent feature. Natural enclosures are planted and at times supported with barbed wire fencing. In some instances, particularly in western Uganda, surveying and registering of the land might follow as individualization has greatly been adopted, depending on the level of development of the community. Enclosures are a common feature in the Ankole area. This is a reflection of livestock policies in Uganda that were established towards ranch development for commercial purposes.

## Problems Affecting Rangeland Areas in Uganda

The problems of rangelands in Uganda and the need for proper management have always existed, at least in latent form, since mankind started to use pastoral land resources in pursuit of various objectives. The main problems of rangelands are those of the scarce, expendable and renewable resources whose quantity and quality can be maintained only through proper management, which ensures high resource productivity both in the short run and in the long run. The nature of rangelands as both expendable and renewable resource and the importance of intergenerational equity considerations posit the need for designing mechanisms to ensure sustainable resource use.

Therangeland situation is complicated in light of two crucial factors. The first being population growth which translates in increased demand for food which traditionally entails opening more land for crops and increased number of livestock. The second factor is related to divergent environmental and conservation policies due to the ever increasing competitive land use. There has been a total failure so far in meeting the increasing demand for food through expansion of cultivated land or extensive livestock. Instead there is increasing pressure on the grazing lands and the natural resources.

Policies related to conservation of natural resources in terms of "protection" such as state forest reserve, games reserves, national parks are subjected to redesigned approach in the form of wise sustainable utilization. There is a need to increase "productivity" to meet the ever increasing food demand coupled with the necessity for sustainable use of natural resources particularly the rangelands. This calls for a sector wide approach to promote an integrated development strategy that stimulates productivity and induces adoption of sustainable resource management practices. The problem is given a special attention by constantly reviewing policies related to land tenure, environment, wetlands, pastoral lands, forest and national parks. As a result serious consideration has been given to policy and institutional options for the management of pastoral lands in dry areas in Uganda.

The management of rangeland resources for sustainable development remains one of the unresolved issues facing policy analysts and development agencies.

Degradation of resources, low productivity, lack of food security, and increasing social tension from more frequent pastoral land resource use conflicts, a declining grazing land area, and relatively low levels of social welfare are persistent concerns for the pastoralists. Colonial and post-independence governments, together with development agencies and non-government organizations (NGOs), have invested substantial monetary resources in efforts to improve pastoral land productivity without commensurate success in sustainable development.

Moreover, development interventions disrupt social institutions and the efficiency of traditional pastoral land management. With population growth and increasing levels of environmental awareness, the issues of sustainable resource use, food security, and social stability have become central to Ugandan national development programs.Conservation strategies have mainly focused on forests and wildlife, taking into account catchment areas, conservation of biodiversity, and areas with a high concentration of rare wildlife species. There is no systematic program to integrate the pastoralist and range resources into such general resource conservation schemes.

As once stated by Sandford (1993) major analyses at both micro and macro levels show that development interventions in pastoral land areas in Africa have failed to generate higher levels of productivity, improve the welfare of local communities, or protect pastoral land from degradation. Land degradation is one of the fundamental issues confronting sub Saharan Africa (SSA) in its efforts to increase agricultural production, reduce poverty and alleviate food insecurity. As population grows and food demand increases while the pastoral land areas are shrinking, the traditional extensive use of land is no longer feasible.

Instead, we have to find the solution that increases "productivity", implying the adoption of intensive farming techniques. Future increases in agricultural production will have to come from yield increases rather than area expansion. However pastoral lands and the pastoral system in particular, pose a more complicated problem than other farming crop and mixed farming systems on merely degraded land other than arid pastoral areas. This is due to the high vulnerability of the pastoral system, which is heavily embedded in strong cultural and historic values.

Besides fragility of pastoral lands, natural hazards such as droughts, floods and wild life issues, these cultural and historical aspects have to be taken into account when designing policies to accelerate technology adoption (Squires, 1998). There is need to examine policy and institutional options that promote accessibility (security of tenure), equity, stability and adoption of improved technologies. While sustainable use strategies are certainly key to environmentally sustainable development, the link between sustainability and development is crucial.

The primary concern of rangeland user is one's immediate survival. Access to pastoral land use under an equitable environment is crucial for survival. Sustainability is a long term undertaking, only attained or seriously considered by the pastoral land user only when the practices pertaining to it facilitate access to and use of the pastoral land with ease for immediate needs. It is crucial to explore and understand in practice what are sustainable strategies and conservation techniques, and who implements or adopts them and for what purpose.

In Uganda the importance of rangelands and the need for their proper management is reflected in their contribution to the livelihood of the people and the importance of livestock in the national economy. The livestock sub-sector contributes about 8 per cent of the country's GDP, 90 per cent of which is produced by small herders, with only 10 per cent deriving from commercial ranches (see Figure 14.5). This production level makes Uganda almost self-sufficient in meat, a position similar to that of the food crop production sub-sector.

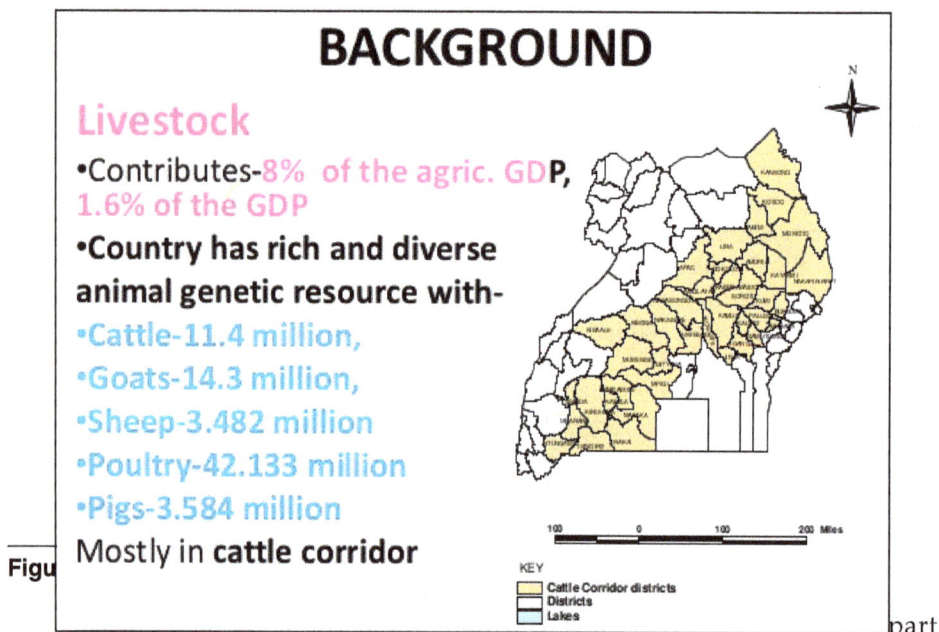

# BACKGROUND

## Livestock

•Contributes-8% of the agric. GDP, 1.6% of the GDP

•**Country has rich and diverse animal genetic resource with-**

•Cattle-11.4 million,

•Goats-14.3 million,

•Sheep-3.482 million

•Poultry-42.133 million

•Pigs-3.584 million

Mostly in **cattle corridor**

KEY
Cattle Corridor districts
Districts
Lakes

Figu...

...part supports increasing livestock populations without improved pasture productivity. Overgrazing as well as under grazing is evident. These result in pasture degradation with change in vegetation cover quantitatively, in terms of biomass; and qualitatively in terms of replacement of high value grass species by unpalatable poor quality grass species and thorny shrubs all of which tend to lead to degradation and promote soil erosion and possibility of gully formation.

Traditional farming systems, which evolved over thousands of years, contained strategies for coping with the unfavorable physical, climatic and biological environment under which people farmed. The coping mechanisms were passive in that people simply adjusted their activities to nature without trying to change the natural situation. Crop farmers practiced shifting cultivation and crop rotations.

While pastoralists developed pastoral systems with various degrees of sophistication, herds were moved continuously following no set pattern along pre-determined routes each year in search of water and pasture following the seasonal rainfall pattern. This practice is equivalent to an "open access" situation with no or minimal institutional development and control. Under this system, the herds moved along pre-determined routes each year in search for water and pasture following the seasonal rainfall pattern.

Pastoralists also kept a diversity of herds to cope with droughts: *e.g.* sheep and goats have high reproductive rates, lactate even in dry periods and goats feed on a wide range of vegetation. They kept the type of animals suitable for the existing environment conditions: disease resistant livestock which could survive under stress of poor grazing conditions, high temperatures and constant movement. But these animals were poor yielders and pastoralists kept great numbers to satisfy their subsistence requirements. The management system was centered on the pastoralists' subsistence needs. While there is a class of elites who are selling and increasing productivity, they are doing so at the expense of those traditional pastoralists who cannot settle and keep on moving along the cattle corridor and beyond the known ancestral grazing grounds.

## Privatization of Communal Rangeland

There has been a high level of privatization of the communal rangelands throughout the entire corridor. The most affected areas are Ntungamo, Mbarara, Rakai, Kiboga, Luwero and even Karamoja. This move has been spearheaded by multiple land users who have settled on previous rangeland and practice crop production alone or in conjunction with livestock keeping. There has been a reduction in the available grazing land on communal land in the areas. The displaced pastoralists graze on the reduced grazing area on the communal grazing land leading to overgrazing and land degradation while the others have either become landless or have moved outside their traditional grazing area, thereby increasing livestock population densities in the new area of invasion with resultant overgrazing and social tension.

The most serious incidence of landlessness of pastoralists as the result of crop farmers' encroachment is in Kasese where the Basongora pastoralists have been completely displaced by the Bakonzo agriculturalists. The incidence of conflicts between the displaced and the encroachers, at the local scene, has reached alarming proportion resulting even in violent encounters. This is the experience in Ntungamo and some parts of Nyabushozi and Karamoja. Conflicts have also risen between old pastoralists and cropper and new comers which have also led to violence. This is the case in Kabarolearea.

## Criticality of People's Participation in Drought Management and Desertification Control

Pastoral people in the rangelands have maximum stake in managing their own affairs. Therefore, lack of their participation has caused growing conflicts between pastoralists and ranchers. This situation has exacerbated drought and desertification without attention from pastoralists living in such disaster areas.

It was established that the introduction of technologies by policy makers without involving resource users (pastoralists) cannot be effective without end users getting involved. For instance, policy makers draft policy interventions without involving people. In the Cattle Corridor, animal sanctuaries and rice scheme have been introduced hindering pastoralists' movement hence causing conflicts.Water dams have been dug and constructed in areas that are not accessible to pastoral populations and enclosed by the elite.

More so, agricultural practices and crops that are drought resistant are introduced without pastoral people's participation. The pastoralists should be educated in modern farming methods so that agricultural output is increased and overgrazing is reduced.

Issues of sustainable resource use, food security, and social stability have become central to Ugandan national development programs. Pastoralists are not trained in taking advantage of emerging opportunities in their areas such as mining and processing gold, coal, diamond and reduce pressure on land and pastures. Therefore participation is crucial to resolve conflict before they occur and to ensure that drought is managed and desertification is controlled using the residents within these areas without compromising the environment and new technologies.

## Capacity Building for People's Participation in Drought Management and Desertification Control

Since colonial times a number of changes have taken place in the management of pastoral land resources in Uganda. The process took the form of sedenterization of some pastoralists who hold the land under communal land tenure system, the creation of privately owned ranches such as the Ankole/Masaka Ranching Scheme, in which the land is held under titled leasehold land tenure, and the creation of various national parks and game reserves located within the cattle corridor. The impact of these development policies varies according to the aridity of the pastoral land. These changes reduced the pastoral land area available to traditional pastoralists, and became a source of conflict, both within the cattle corridor and also with neighboring districts.

A National Action Plan to Combat Drought and Desertification was put in place. The Action Plan calls, *inter alia*, for education measures and raising awareness on desertification issues. Farmers and herdsmen are encouraged to diversify economic activities and find alternative livelihoods. However, rangeland management degenerated into "open access" leading to overgrazing, pasture degradation and soil erosion. Ugandan pastoral lands display a high degree of pasture and soil degradation and some parts have traces of desertification. Most crucially these rangeland management practices fail short of recognizing pastoralism (livestock farming) as an economic activity that needs business, skills and quick decision making mechanisms by the 'farmer'. For a long time the government has neglected policies that would promote investment in the management of pastoral lands.

# Interventions for Capacity Building for Drought Control and Desertification in Pastoral Areas

## Knowledge Upgradation

The pastoralists should be educated in modern farming methods so that agricultural output is increased. They ought to be trained in taking advantage of emerging opportunities in their areas such as mining and processing gold, coal, and diamond. Education should also be utilized so that more pastoral young people are released into other professions such medicine, teaching and engineering.

Likewise, pastoralists should further be trained in modern ways of farming such as using high yielding crops that ensure optimal production. This will reduce tensions in pastoral areas, stress on pastures and getting more money from high yield crops.

## Agricultural Modernization in all Fields

The government has restructured the civil service, decentralized governance through the Decentralization Program, embarked on a universal primary education program under UPE program and is implementing land tenure reforms. Modernization has in principle three basic elements:

* ★ Development: appropriate technologies and technological innovations should be introduced and adopted in production and marketing in the broadest sense
* ★ Planning should be based on scientific principles and modern methodologies

There should be human development both at the individual and the societal levels. The relevance of modernization in resource management is reflected in technology and technological innovations in terms of research, water supply systems, design of integrated farming systems, pests and disease control methods, new marketing systems, formation of new farming systems and human development.

Institutional framework should be strengthened to manage communal land where elders should be involved and be given more powers to teach pastoralists land management in response to drought and desertification

The pastoralists should de-stock in anticipation of drought. At the same time, early warning adjustment technologies should be used locally by pastoralists in order to detect drought early and allow pastoralists enough preparation to move to less affected areas.

Drought resistant trees such as acacia should be planted in the Cattle Corridor to provide shade and moderate air temperature and at the same time provide wood and Gum Arabic for income generation. Gum Arabic is more harvested in more severe droughts.

Simple technology transfer should be used by pastoralists to enhance their productivity in the rangelands. Currently, such technology is only utilized by only policy makers.

Given mobility of pastoralists, there is need to resolve conflicts that arise from their grazing in the privatized land which they still continue to regard as communal.

Though transhumance is no longer viable, pastoralism should be enhanced through transfer of cattle by Lorries rather than by herding because individualization has blocked free and communal movement of cattle.

Pastoralists should produce more products of national and international standards for markets such as cheese, packed milk *etc.* in order to reduce focus on big numbers of cattle hence reducing overgrazing.

Infrastructure in pastoral areas should be developed such as schools, factories and hospitals so that pastoralists concentrate on developmental activities.

A more holistic approach should be used where different land users and stakeholders (pastoralists, charcoal burners, agriculturalists, and wildlife *etc.*) are brought together to identify problems and solve them together.Agro-pastoralism should be encouraged so that pastoralists settle and manage land sustainably.

**Provision of Water**

As stated earlier the main innovations include the construction of watering points, in form of dams in Mbarara, Ibanda, Kiruhura,Rakai, Sembabule and Karamoja in order to cut down the Karamojongs' nomadic practices to neighboring districts and cattle rustling. There are clear signs of overgrazing and serious soil erosion along the trek routes to and around the new watering points, because the provision of more water has encouraged the growth of livestock beyond the number supportable by the available pasture (Goldschmidt, 1981).

**Sedentarization of Nomads through Privatization of Pastoral Land and Ranche**

* The government has established ranches in the South West under the Masaka/Ankole ranching scheme where original big ranches of (5sq. miles) have been restructured into smaller ones (1 sq.mile). The second is the newly formed ranching schemes in Kiboga district covering 30 sq. miles.

* Several private ranches are being established on public land under leasehold arrangements in Luwero, Kiboga,Nakasongora and Masindi districts.

* The Land Act of 1998 provided for the formation of group ranches in addition to ensuring security of tenure for tenants under all types of land tenure systems.

* A high level of spontaneous individualization of communal land by pastoralists and crop farmers is going on in the Cattle Corridorwith tacit support of the government which has however resulted in displacement of pastoralists and cause conflicts among local communities and areas.

* Construction of watering points, as the dams in Mbarara, Ibanda, Kiruhura,Rakai, Sembabule, and similar constructions planned for Karamoja aimed at cutting down nomadic practices to neighboring districts and cattle rustling.

## Conclusions

The study has established that there is a gross involvement of pastoral communities in solving the problems affecting the rangeland economy in Uganda, more especially in the area popularly known as the Cattle Corridor. The findings have shown that various institutions working on capacity building and people's participation in drought management and desertification control in rangelandareasin Uganda have not involved people in resolving conflicts related to pasture in their areas.

In the light of institutional problems among pastoral communities in the rangeland areas, a participatory approach involving local communities is recommended in determining the best opportunities available for the management of pastoral lands. This promotes ownership and captures their indigenous knowledge. To improve rangeland management for sustainable development and improve welfare of the local communities, there is, among others, a need to increase production and productivity of rangeland through increased off-take of livestock and livestock products, and educate the pastoralists to take advantage of emerging opportunities in the economy.

Further interventions would include enhanced sedentarization of nomads through privatization of rangeland and creation of ranches for them. This would entail application of appropriate technologies to enhance the utilization of rangeland resources with particular reference to the use of water and drought-resistant pasture and crops. The establishment of a functional livestock market system may go a long way in empowering pastoralists to acquire land and settle to increase productivity and production. There is also a need to establish an institutional framework to ensure that local actors participate in designing and implementing various policies and programs in the rangelands.

## References

1.  American Institute of biological Sciences, (2004). Analysis Fingers Causes Of Desertification; BioScience Press Release September; American Institute of Biological Sciences; Washington DC.

2.  Bhavnani E., Birkmann, J., Thywissen, K., Renaud, F., Sakulski, D., Affeltranger, B., Shen, X. (2008). Climate Change, a Hazard to Human Security; Human Security and Climate Change; An International Workshop, Oslo, Norway, 21–23 June

3.  Esikuri, T. (2005). International Conventions and Environmental Legislation and their Applications.Desertification A review of the concept, *Encyclopaedia of Climatology*, J.E. Oliver and R. Fairbridge, eds., Hutchinson. Ross Publishing Company.

4.  Gautam, M. (2006). Managing Drought in Sub-Saharan Africa: policy perspectives. Invited Paper Prepared for a Panel Session on Drought: Economic Consequences and Policies for Mitigation, at the IAAE Conference, Gold Coast, Queensland, Australia, August 12–18, 2006.

5.  Goldschmidt, G. (1981). Environmental Change and Migration; Global warming and the Third World; Tiempo - Issue 42, December

6.  Hellmuth, L., Johnson, Keier K., Mlcom, B. (2007).The Influence of Climate Variability on Pastoral Conflict in Africa; Human Security and Climate Change An International Workshop Oslo, Norway, 21–23 June

7.  Holden and Shiferaw (2004)."Human Vulnerability, Dislocation and Resettlement: Adaptation Processes of River-bank Erosion-induced Displacees in Africa. *Disasters*. 28(1): 41

8.  Hulme, M. and Kelly M. (2013).Exploring the links between desertification and climate change. *Environment* 35: 4,39-11,45.

9.  KisambaMugerwa, (1992). Rangeland Tenure And Resource Management: An Overview Of Pastoralism In Uganda. Research Paper Prepared for Makerere Institute of Social Research and the Land Tenure Center, January.

10. Riebsame *et al.* (2005). Drought-Related Conflicts, Management and Resolution in the West African Sahel; Human Security and Climate Change An International Workshop Oslo, Norway, 21–23 June.

11. RosegrantL., Lindley, W.I., Bruening, T.H. and Doron, N. (2002). Agricultural Education for Sustainable Rural Development: Challenges for Developing Countries in the 21st Century; Extension, Education and Communication Service (SDRE); FAO Research; Rome, Italy.

12. Rowell, N, Polkion, J, Hulk, V, Doernsk, L. (2012). Climate Change, Population Drift and Violent Conflict over Land Resources in Africa; Human Security and Climate Change An International Workshop Oslo, Norway, 21–23 June.

13. Sandford, H. (1993). Towards Global Food Security: Fighting against hunger. Towards Earth Summit, Social Briefing No.2.

14. Squires, M. (1998). Food security options in Zimbabwe: multiple threats, multiple opportunities?' Country Food Security Options Paper No. 5, Forum for Food Security in Southern Africa.

15. Uganda Bureau of Statistics (UBOS) (2016). Statistical Abstract, October.

16. UNCCD (2004). Ten years on: UN marks World Day to Combat Desertification; Observances worldwide on June 17.

17. UN Department of Economic and Social Affairs (2016). New York.

18. UNEP. (1994). United Nations Convention to Combat Desertification in those countries experiencing drought and/or desertification, particularly in Africa, Nairobi.

19. Yang, J. and Prince, S.D. (2010). Remote sensing of savanna vegetation changes in eastern Zambia 1972-1989. *International Journal of Remote Sensing* 21: 301-332.

20. White Paper on Rangeland in Uganda (2002). Government of Uganda, Kampala.

21. Kisamba-Mugerwa (2001). *Rangelands Management Policy in Uganda*. A Paper Prepared for the International Conference on Policy and Institutional Options for the Management of Rangelands in dry Areas May 7 – 11 (Hammamet, Tunisia).

## Mashhad Resolution on

# Drought Management and Desertification Control

**WE, THE DELEGATES** to the 3-days International Workshop on "Drought Management and Desertification Control", jointly organised by the Centre for Science and Technology of the Non-Aligned and Other Developing Countries (NAM S&T Centre) and Ferdowsi University of Mashhad, and Khorasan Razavi Agricultural and Natural Resources Research and Education Center (KRANRREC), at Mashhad, Iran during 22nd– 24th May 2017, comprising scientists, researchers, academicians and policy makers from Bangladesh, Bhutan, Cuba, India, Indonesia, Iran, Iraq, Malaysia, Mauritius, Myanmar, Nepal, Nigeria, Palestine, Sri Lanka and Uganda;

**RECOGNIZING** that though drought, desertification and land degradation in dry lands have been under focus at national and international levels for the past four decades and that some reclamation efforts have also followed, the problem has persisted and even aggravated in some respects;

**HAVING** considered that the consequences of these environmental issues and the affected populations are more serious particularly in the developing world, there is paucity of enabling resources, and/or inadequacy of expertise that constrain effective desertification control and development efforts;

**HAVING** deliberated upon the experience over the years, it is seen that the problem of land degradation cannot be tackled by just physical remedies of one or the other types alone and the underlying driving factors namely, the socio-economic milieu and the entailing exploitative systems need an equal attention;

**TAKING** into account that manifestations of desertification and drought are strongly inter-linked and that major accentuation of the adverse consequences and land degradation and human strife occur during prolonged periods of deficit rainfall or drought, planning and development strategies need to have an inbuilt recognition and preparedness for such eventualities;

**UNANIMOUSLY RESOLVE THAT:** The Mashhad International Workshop has helped in discussing issues and strategies relevant to enhancement of land productivity on sustainable basis in some of the affected countries with varied socio-cultural levels of economy and nature of biophysical resources to mutual advantage;

### AND RECOMMEND THE FOLLOWING

★ Since the cost and effort involved in land degradation control are much higher than the prevention, early recognition of the problem and matching conservation-oriented management effort should form a cornerstone of long-term plans of action.

★ The recent driving cause underlying all natural resources degradation problems in most countries is the pressure inducted by human activities. Even with possible resource conservation and development effort, the land production may be inadequate to meet the current and aspirational needs. Therefore, as a long term strategy it is important to generate alternative, non-farm employment potential in different sectors of economy.

★ Experience over the decades has shown that though technologies have worked in restoring degraded lands, the resultant gains have got undermined to a considerable measure by inadequate management regime arising from socio-economic disparities and community disharmony.

★ The problem of safeguarding ecological integrity/services and ensuring livelihood security are constrained by inadequacy of required resource base and capacity and hence a need does exist for a stronger international cooperation amongst the affected and other nations across the world.

★ Evaluations have shown that government policies and programs for rehabilitation of degraded lands have not been integrated adequately with the concerns and participation of local stakeholders. Immense amount of emphasis should be placed on the concerns, capacity limitations and expectations of the directly affected populations.

★ Climate change and global warming severely impact the performance of agricultural and natural resources sectors and hence adaptation and mitigation strategies are emphasised.

★ The time has come that developing countries make rigorous and scientifically sound analysis of all the mitigation and adaptation efforts. Such efforts will come very handy in future course of action and in experience-sharing workshops at the forthcoming Rio+20 and UNCCD international activities.

★ Training, education and capacity building are essential components of the strategy for effective desertification control.

★ Indigenous knowledge reflects a huge understanding of potentialities and thus should be appropriately adopted as an important component of resource-constrained regions management and development.

★ Non-Governmental Organizations can play a critical role in capacity development and in this regard should be appropriately involved and provided necessary support in various programs and projects.

★ Research and new technologies are crucial in the mitigation and adaptation of the impact of climate change and desertification and should be included in all projects.

★ Drought and desertification are global issues and hence involvement of all countries in management and control is seriously needed.

THUS RESOLVED AND ADOPTED ON THE 23$^{RD}$ MAY 2017 AT MASHHAD, I.R. IRAN.

www.ingramcontent.com/pod-product-compliance
Lightning Source LLC
Chambersburg PA
CBHW050517190326
41458CB00005B/1567